Genetic, Linguistic and Archaeological Perspectives on Human Diversity in Southeast Asia

Recent Advances in Human Biology

Series Editor: Charles E. Oxnard *(University of Western Australia, Australia)*

Published

Vol. 10 *Photoreceptor Cell Biology and Inherited Retinal Degenerations*
edited by David S. Williams

Vol. 9 *The Ubiquitin-Proteasome Proteolytic System: From Classical Biochemistry to Human Diseases*
edited by Aaron J. Ciechanover & Maria G. Masucci

Vol. 8 *Genetic, Linguistic and Archaeological Perspectives on Human Diversity in Southeast Asia*
edited by Li Jin, Mark Seielstad & Chunjie Xiao

Vol. 7 *Morphometrics for the Life Sciences*
by Pete E. Lestrel

Vol. 4 *The Natural History of the Doucs and Snub-Nosed Monkeys*
edited by Nina G. Jablonski

Vol. 3 *The Origins and Past of Modern Humans: Towards Reconciliation*
edited by Omoto Keiichi et al.

Vol. 2 *Bone Structure and Remodeling: Proceedings of the Symposium held at the Second World Congress of Biomechanics*
edited by Odgaards A. et al.

Vol. 1 *The Origin and Past of Modern Humans as Viewed from DNA: Proceedings of the Workshop on the Origin and Past of Homo Sapiens Sapiens as Viewed from DNA — Theoretical Approach*
edited by Brenner Sydney & Hanihara Kazuro

Recent Advances in Human Biology – Volume 8

Series editor: Charles E. Oxnard
Centre for Human Biology
The University of Western Australia

Genetic, Linguistic and Archaeological Perspectives on Human Diversity in Southeast Asia

Yunnan University, China 26 – 27 June 2000

Editors

Li Jin
Human Genetics Center, School of Public Health
University of Texas, Houston, USA

Mark Seielstad
Program for Population Genetics, Harvard School of Public Health
Harvard University, Boston, USA

Chunjie Xiao
Human Genetics Center, Yunnan University, China

World Scientific
New Jersey • London • Singapore • Hong Kong

Published by

World Scientific Publishing Co. Pte. Ltd.
5 Toh Tuck Link, Singapore 596224
USA office: 27 Warren Street, Suite 401-402, Hackensack, NJ 07601
UK office: 57 Shelton Street, Covent Garden, London WC2H 9HE

British Library Cataloguing-in-Publication Data
A catalogue record for this book is available from the British Library.

Recent Advances in Human Biology — Vol. 8
GENETIC, LINGUISTIC AND ARCHAEOLOGICAL PERSPECTIVES ON HUMAN DIVERSITY IN SOUTHEAST ASIA

ISBN-13 978-981-02-4784-3
ISBN-10 981-02-4784-2

PREFACE

Southeast Asia has the longest record of human (or hominid) habitation outside of Africa, and it has occupied a pivotal role for most of the nearly two million years following the arrival of *Homo erectus* in what is now Java. The first modern occupants of Oceania (or Sahuland, generally, including Australia), East Asia, Polynesia, Madagascar, and perhaps even the Americas emanated directly from this region, which itself has been the scene of tremendous cultural innovation and variation.

Despite the apparent centrality of the region in the history of our species, it has been relatively little studied by geneticists. This situation is rapidly changing, with the arrival of new generations of scientists in the region; an increasing awareness of its cultural and genetic variation amongst scientists in East Asia, Europe, and the Americas; and the arrival of new and powerful genetic markers, mostly from the Y chromosome and mitochondrial DNA. With a view toward galvanizing research in this area and creating a collaborative and cooperative spirit amongst scientists interested in the region, we organized an international conference in June 2000 at the University of Yunnan in Kunming, China. The papers in the present volume result from this intimate meeting, though papers from Satoshi Horai (Hayama; mitochondrial variation in East Asian populations) Jeremy Martinson (Nottingham; hemoglobin variants in Insular SE Asia), Sangkot Marzuki (Jakarta; mitochondrial variation in Insular SE Asia), Laurent Sagart (Paris; linguistics), and Mark Stoneking (Leipzig; evidence against recombination in mitochondrial DNA) are not published here.

At the outset, we identified five major lines of research that are or might be the subject of population genetics research in the region:

1. Assessing the probability of any gene flow between *Homo erectus* and modern human populations in SE Asia. Currently, there is no evidence of any such gene flow, but the possibility of limited genetic contact has not been entirely excluded. Since *Homo erectus* appears to have occupied the region until as recently as 30,000 years ago, this is one region where modern and archaic humans may have overlapped for the greatest length of time. Recent work has all but excluded the possibility of Y-chromosomal gene flow from archaic to modern populations in Asia, but the autosomes with their fourfold

greater probability of preserving an ancient lineage have not yet been thoroughly examined.

2. Tracing and dating the entry and subsequent migrations of the first anatomically modern humans, who appear to have followed a coastal route from Africa, reaching SE Asia more rapidly than points further inland.

3. Assessing the possibility of an agriculturally driven demic (*vs.* cultural) expansion of more northern populations into SE Asia, possibly at the expense of 'relict' populations such as the Negritos of Malaysia and the Philippines.

4. Determining the contribution of Asian populations to the settlement of the Pacific and the Americas.

5. Assessing the effects of cultural practices such as polygyny, patrilocality, and uxorilocality on the patterns of variation on the X and Y chromosome, the autosomes, and mitochondrial DNA.

Population Genetics is a synthetic field, which must integrate the genetic results with those from archeology, paleontology, and linguistics. The first chapters present the perspectives of linguists, archeologists, and paleontologists on the origins of East and Southeast Asian populations, and they are followed by sections on genetic perspectives on the 'peopling' of each of the major geographic regions whose beginnings may trace to SE Asia.

This international conference was a great success in bringing together a considerable number of researchers interested in SE Asian populations. We hope that this small community of researchers will continue the spirit of collaboration begun at this conference, placing SE Asian genetic studies at the forefront of population genetics research.

We would like to thank the staff of the Biology Department at Yunnan University for their unflagging efforts to make this conference a success. Support for the attendance of some participants was provided by Mike Pellini and Kristin Ardlie of Genomics Collaborative in Cambridge, Massachusetts. Sara Ann Barton of the Human Genetics Center, University of Texas, Houston worked tirelessly and extremely thoroughly to edit and typeset this entire volume. Her efforts are greatly appreciated and the results are clear in the book's pleasing layout.

Mark Seielstad
Cascade, Idaho
Li Jin
Houston, Texas
Chunjie Xiao
Kunming, Yunnan

CONTRIBUTORS

Chu, Jiayou
Institute of Medical Biology
Chinese Academy of Medical Sciences
379 Jiaoling Rd.
Kunming 650118, P.R. China
E-mail: chujy@public.km.yn.cn

Deka, Ranjan
Department of Environmental Sciences
University of Cincinnati
P.O. Box 670056
Cincinnati, Ohio 45267-0056, USA
Fax: 513-558-4397
E-mail: ranjandeka@uc.edu

Higham, C.F.W.
University of Otago
P.O. Box 56
Dunedin, New Zealand
Fax: 0064-3-479-9095
E-mail: c_higham@yahoo.com

Jin, Li
Center for Genome Information
Department of Environmental Health
Kettering 251
3223 Eden Ave.
University of Cincinnati College of Medicine
Cincinnati, OH 45267-0056, USA
Fax: (513) 558-0071
E-mail: li.jin@uc.edu

Kangwanpong, Daoroong
Department of Biology
Chiang Mai University
Chiang Mai 50200, Thailand
E-mail: scidkngw@chiangmai.ac.th

Lum, J. Koji
Department of International Affairs and Tropical Medicine
Tokyo Women's Medical University
8-1 Kawada-cho, Shinjuku-ku
Tokyo 162-8666, Japan
Fax/Tel: 81-35-269-7422
E-mail: jekl@research.twmu.ac.jp
i8uni@yahoo.com

Roseman, Charles C.
Department of Anthropological Sciences
Stanford University
Stanford, California 94305-2117, USA
Fax: 650-725-9996
E-mail: croseman@stanford.edu

Seielstad, Mark
Program for Population Genetics
Harvard School of Public Health
665 Huntington Ave.
Boston, Massachusetts 02115-6096, USA
Fax: 617-432-2956
E-mail: mark@ppg.harvard.edu

Singh, Nadia
Program for Population Genetics
Harvard School of Public Health
665 Huntington Ave.
Boston, Massachusetts 02115-6096, USA
E-mail: ndsingh@stanford.edu

Srikummool, Metawee
Department of Biology
Chiang Mai University
Chiang Mai 50200, Thailand

Su, Bing
Human Genetics Center
School of Public Health
University of Texas
P.O. Box 20186
Houston, Texas 77224, USA
Fax: 713-500-0900
E-mail: subn@ucmail.uc.edu

Tan, S.G.
Department of Biology
Faculty of Science and Environmental Studies
Universiti Putra Malaysia
43400 UPM Serdang, Malaysia
Fax: 603-865-67454
E-mail: sgtan@fsas.upm.edu.my

Thosarat, Rachanie
Fine Arts Department
9th Regional Office of Archaeology and National Museum
Phimai, Nakhon Ratchasima
Thailand 30110
E-mail: fad9@loxinfo.co.th

Underhill, Peter A.
Department of Genetics
Stanford University
300 Pasteur Drive
Stanford, California 94305-5120, USA
Fax: 650-725-1534
E-mail: under@stanford.edu

Wang, S.-Y.
Department of Electronic Engineering
City University of Hong Kong
Kowloon, Hong Kong, China
Fax: 852-2788-7791
E-mail: eewsyw@cityu.edu.hk

Xiao, Chunjie
Human Genetics Center
Yunnan University
Kunming, 650091Yunnan, China
E-mail: cjxiao@public.km.yn.cn

CONTENTS

Part I: Prehistory of Human Populations, Archaeological, Linguistic and Paleontological Perspectives

PREHISTORY, LANGUAGE AND HUMAN BIOLOGY: IS THERE A CONSENSUS IN EAST AND SOUTHEAST ASIA?

C.F.W. HIGHAM

University of Otago

Integrating archaeology, historic linguistics and genetics when investigating major trends in world prehistory is one of the most alluring, but at the same time, one of the most testing of objectives. One only has to review some of the writings on prehistory by linguists or geneticists, to be cautious of making similar pronouncements in reverse. Circular arguments are easily entered into where disciplines coincide. Thus a possibility advanced by a linguist may be taken to support an archaeological overview, only for the latter to be adopted with enthusiasm by the same linguist in support of the original proposition. Again, a linguist might identify a split between two proto-languages, and make a reasoned guess on when this occurred, only to find that same guess adopted as valid by a prehistorian. The logical resolution to this situation is to take results of each discipline separately when treating the same issue, and then investigate any signs of congruence. In this manner, there is likely to be a fertile relationship, in which hypotheses based on one source of information can be tested on the basis of another. This is elegantly to be seen in Pawley's linguistic reconstruction of what a Polynesian house would have looked like, to be confirmed in due course by Green (Green and Pawley in press) through archaeological excavation.

This paper will adopt this approach in considering one of the six themes covered by this book, and two others more briefly.

1. Tracing and dating the entry and subsequent migrations of the first anatomically modern humans

Archaeology still has little to offer this issue. Ideally, there should be archaeological evidence for the arrival of anatomically modern humans, as there is in Europe, for example, in terms of material culture. In Thailand, there is a void between convincing evidence for *Homo erectus* and late Pleistocene hunter-gatherer groups dating from about 38,000 years ago. Yet the settlement of Australia would require the establishment of ancestral hunter-gatherers on the Asian mainland well before that date. Excavations at Lang Rongrien, in Krabi Province of Thailand, have provided the longest and most important hunter-gatherer sequence available on the mainland of Southeast Asia. Radiocarbon determinations from the lower contexts of Anderson's excavations in 1983 indicated settlement between 38,000-27,000 years ago. These lower layers at Lang Rongrien were only found after Anderson determined to remove what looked like the stony bedrock of the cavern, only to find that it had formed through roof fall. Underneath lay the remains of early occupation. It is highly likely that similar material awaits discovery elsewhere. During this period, corresponding to the extreme cold of the late Pleistocene in higher latitudes, the sea level was much lower than at present, and the coast was located at least 30 km from the cave. Periodic but brief visits by hunter-gatherers have left the remains of their hearths, the animals they hunted, and their stone tools. The local chert was used for manufacturing flaked knives, scrapers and chopping tools.

There is also a late Pleistocene tradition of hunting and gathering in northern Vietnam known as the Son Vi culture. (*c.* 18000--9000 BC). Here, and elsewhere in the uplands of Southeast Asia, it is followed by the well-known Hoabinhian. The typical stone tools of this group come from thin occupation layers in small rock shelters. Biological remains reveal subsistence based on hunting the forest fauna and exploiting streams and rivers for fish and shellfish. Plant remains include a wide variety of food plants as well as those which supply poisons and stimulants. Very few sites include human remains, although at Lang Cao, Colani (1927) found no fewer than 200 skulls within an area of only 25 square meters, propped up by stones, with few associated limb bones.

Such inland hunter-gatherers may well have been minutely adapted to foraging in a forested habitat, but present only one side of the picture. The tropical shore is a very rich habitat, and coastal hunter-gatherers present a picture of social complexity in sharp contrast with those of the interior. Here, we find large settlements, which incorporate inhumation cemeteries in which the dead were interred with a range of grave items. Rich food resources encouraged a degree of permanence, which fostered a ceramic industry, and development of a sophisticated technological repertoire based on stone, bone and shell. But there is a further void in our understanding of maritime hunter-gatherers. The rising Holocene Sea covered huge areas of lowland Southeast Asia, drowning early settlement sites. Only when the sea rose higher than its present level can we trace hunter-gatherer sites on old raised shorelines.

The time depth for these hunter-gatherers, particularly in the early stages of any alleged intrusion by anatomically modern people, rules out any consideration of historic linguistics. In terms of human biology, however, there is much untapped potential. As Bulbeck (1999) has shown, surviving groups of hunter-gatherers in the Philippines, southern Cambodia, the Malay peninsula and the Andaman Islands have in common a particularly dark skin, short stature and woolly hair. Most live in a tropical rainforest habitat, presumably after being marginalised by more powerful agricultural societies. The mean Andamanese stature is only 1.46 m for males and 1.38 m for females. Their language is virtually unstudied, but comprises one of only 24 phyla recognized worldwide (Blench and Spriggs 1998). An insular isolation has ensured the survival of their language, but this is not the case in peninsular Malaysia, where local hunter-gatherers have adopted an Austroasiatic language. Surviving Negrito hunter-gatherers, from the Andamans to the Philippines and from the Aborigines of the Atherton Tableland to Tasmania, have encouraged numerous explanatory hypotheses. Bellwood (1997), for example, has suggested that those living in the peninsula Malaysia might be surviving groups from the local late Hoabinhian hunter gatherers documented archaeologically at such sites as Gua Cha and Lang Rongrien. Bulbeck (1999) however, has suggested that they represent an adaptation to the rainforest habitat, which favors small body size, while woolly hair is adaptive to heat and water proofs the head. In terms of craniometry, he notes parallels between the Andamanese and sub-Saharan groups in Africa, while dental morphology again points in the general direction of Africa. He concludes by suggesting that there was expansion into the area immediately from South Asia but ultimately from Africa, about 100,000 years ago.

This proposal can be advanced on the basis of the DNA analyses on two fronts. The first is to characterize the DNA sequences of the Andamanese and other groups of Negritos and then compare them with, for example, Southern Mongoloid and African populations. During the early 20th century, Radcliffe-Brown collected hair samples from Andamanese and these have been retained in Cambridge. Collected prior to any major gene flow with other groups, the analysis of these samples is vital. Early indications are that they reveal close affinities with African sequences (Hagelberg pers. comm. 2000). The second avenue is to sequence prehistoric bone samples from Hoabinhian and coastal hunter-gatherer groups. This is a challenge for the future.

2. Assessing the possibility of an agriculturally driven demic (vs. cultural) expansion of northern populations into SE Asia, at the possible expense of 'relict' populations such as the Negritos

Identifying and explaining the origins of agriculture is a key issue in East and Southeast Asian prehistory. Archaeological documentation and validation of the transition from hunting and gathering to agriculture is the necessary first step, before the implications in terms of expanding human settlement are sought. At the same time, while archaeology proceeds at its own stately pace, historic linguists are themselves addressing the same issues but relatively at the speed of light. This is particularly well illustrated by Van Driem's (1998) suggestion that the establishment of the first millet farmers of the Yellow River basin, the Peiligang and Cishan cultures, represent an intrusion from Sichuan. This might seem possible on linguistic evidence, but has no archaeological validation in Sichuan itself.

The origins and implications of rice domestication are not only vital for understanding East Asian prehistory on its own terms. They also present a unique opportunity to test a model, which has attracted much interest and controversy over the past 15 years. Ammerman and Cavalli-Sforza (1984) and later Renfrew (1987) ignited this issue by suggesting that the Neolithic revolution in the Levant stimulated population growth and expansion which most easily explains the present distribution in the Old World, of Indo-European languages. Since the only other such transition occurred in East Asia, one can inquire if it had similar effects in terms of expansive agricultural societies and language distributions. In this context, we are particularly fortunate that at last, a sequence encompassing this transition has been identified.

The sequence can only be understood with reference to environmental change. The climate of the Yangzi Valley underwent a series of profound changes incorporating the end of the Pleistocene ice age and the oscillations in temperature and rainfall, which characterized the early Holocene. Recent evaluations of pollen spectra and faunal assemblages there reveal a climate 4-10 degrees cooler and much drier between 20,000 and 15,000 BC (Higham and Lu 1998). Rainfall was probably 1000 mm per annum below its present level of 1600 mm, accounting for the predominance of drought resistant plants in the pollen spectra. From 15,000 to about 13,700 years BP, the climate moderated, encouraging the spread of oak and pine, elm and willow but thereafter, and until 10,000 BP, there was a reversal to cold conditions described across Eurasia as the Younger Dryas period. Thereafter, it again became warmer and moister. Broad-leaved trees colonized the valley and the fauna became subtropical. Increased rainfall fed rivers and lakes, and wild rice spread from refugia.

It is against this environmental kaleidoscope that we can measure the significance of recent finds from deep excavations in the caves, that fringe the lacustrine lowlands. The sequence at Diaotonghuan, for example, spans the later Pleistocene into the early Neolithic period. This cavern overlooks a small, swampy basin (Zhao 1998). The excavators have identified 16 sequential zones of occupation, and recovered samples of rice phytoliths, the hard silica bodies found in the rice plant. There was, for example, a surge in the numbers of rice glume phytoliths in zone G, which is tentatively dated to the terminal Pleistocene. These are seen as evidence for the collection of wild rice during the mild phase, which characterized that period. Rice phytoliths were extremely rare during zone F, which corresponds to the Younger Dryas cold phase. However, rice was again abundantly represented during zone E, which is thought to date between 10,000-8,000 BP. About half the sample conforms to a domestic variety of rice. This context also provided the first evidence for pottery in the form of very crude, sometimes cord-marked vessels, which could well have been made in order to cook rice.

A lack of reliable radiocarbon dates makes this a tentative framework, but it gains support from similar sequences in other caves. Xianrendong is located only 800 m from Diaotonghuan, and again has a lower Palaeolithic occupation under a Neolithic horizon containing rice phytoliths. Yuchanyan also overlooks low-lying wetlands, and has provided a sample of fish, turtle and mammalian bone as well as rice husks said to be transitional to the domestic form. Potsherds from this site are dated in the vicinity of 12,500 BP (Yuan and Zhang 1999). Bashidang is a village site, which covers about three hectares. Its lower layers date to about 8000 BP, and excavations in 1993-7 uncovered waterlogged deposits, which had preserved over 15,000 rice grains. These have been ascribed to a cultivated variety (Pei 1998). Water caltrop and lotus, both of which can easily be propagated in marshes and lakes, were also abundantly represented in this settlement, together with hunted and probably domestic animals, pottery vessels, wooden spades and pestles, the foundations of pile dwellings and over 100 human burials. Bashidang is similar in many respects to the settlement of Pengtoushan, found only 20 km to the southwest. Here, we encounter a cemetery in which the dead were interred with complete pottery vessels and exotic stone ornaments. The clay used for making pots was tempered with rice chaff.

Again, there are the remains of houses and every sign of a successful adaptation to the rich resources offered in the middle Yangzi Lakeland. Two radiocarbon determinations obtained from the rice used as a tempering agent are 6420-6990 and 5780-6380 BC (calibrated, Crawford and Chen 1998).

This accumulating body of evidence indicates that the Yangzi Valley was one of the very few areas in Eurasia that witnessed a Neolithic Revolution, the transition from hunting and gathering to agriculture. Population growth is a recurrent characteristic of sedentary agricultural communities. As settlements grow, there is a strong incentive for a segment to move and found a new community. This appears to have followed the establishment of such sites as Pengtoushan and Bashidang. Fenshanbao, which was occupied within the period 8000-7500 BP, lies east of Lake Dongting, and excavations have revealed 50 burials and pottery tempered with rice. To the west, we find agriculture spreading upstream to Chengbeixi in the Three Gorges. In an easterly direction, the famous site of Hemudu was a base for lakeside rice cultivation by 7000 BP.

This sequence has a strong bearing on the Neolithic settlement of Southeast Asia, because it is now possible to trace the expansion of agricultural communities progressively further to the south. Several rivers provide access from the Yangzi Valley to the rich hot lowlands of Lingnan. The Gan and Xiang flow north to Lakes Poyang and Dongting, while the Bei flows south. The first evidence we have for the establishment of rice farmers is, not unexpectedly, in the headwaters of this last river, where the sites Shixia, Xincun, Chuangbanling and Niling date from the early third millennium BC. Shixia in its earliest phase, included a cemetery in which grave goods included jade *cong* (tubes) of deep ritual significance in the Liangzhu culture to the north, as well as bracelets, pendants and split rings. The subsequent Nianyuzhuan culture sites reflect a further spread of agricultural settlement, but began to encounter and interact with rich hunter-gatherer groups commanding the delta of the Zhu River.

The Bei is just one of the rivers which ultimately connects the Yangzi Valley with Southeast Asia. In general, they flow south and radiate out from a hub in the eastern Himalayan foothills. From east to west, they include the Red, Mekong and Chao Phraya systems. Further to the west, this configuration is repeated in the form of the Irrawaddy, Chindwin and Brahmaputra rivers. Given the dense canopied forests that would then have dominated the lowlands of Southeast Asia, the rivers were the principle arteries for communication and movement.

Yunnan is a key area for documenting any expansionary movement of this nature, because it has links with the Yangzi, the Mekong and the Red rivers. Baiyangcun is a site which lies within striking distance of all three. It has a deep stratigraphic sequence, involving over four meters of accumulated cultural material. The initial settlement has been dated to between 2400-2100 BC, and excavations over an area of 225 square meters have revealed the remains of eleven houses and a cemetery. Many of the human remains were found with no cranium, and grave goods were also absent, but the pottery from this phase was decorated with a distinctive series of patterns, incorporating parallel incised lines infilled with impressions (YPM 1981). The nearby site of Dadunzi is rather later, the single radiocarbon date suggesting a mid second millennium BC occupation. Again, house plans were noted, often superimposed over earlier structures, and 27 burials were encountered. Adults were buried in extended positions with no preferred orientation, and infants were interred in mortuary jars. The style of pottery decoration matched that found earlier at Baiyangcun.

Archaeological research in the major river valleys of Southeast Asia has revealed a compelling pattern in which new agricultural villages were established between 2500-2000 BC. In the Red River valley, this phase is seen in many sites of the Phung Nguyen culture. In the Mekong catchment, we find Neolithic phases of occupation at Ban Chiang, Non Kao Noi and Ban Lum Khao and in the valley of the Chao Phraya River, Ban Kao, Non Pa Wai and Ban Tha Kae indicate settlement towards the end of the third millennium BC. A common inhumation burial ritual, domestic pigs, cattle and dogs, and a similar technique of decorating pottery vessels link these sites. They represent a classical case of expansion by intrusive Neolithic people, comparable with the Linear Bandkeramik sites of the European loess lands.

As Bellwood (1997) and Spriggs (1998) have shown, a similar process of expansion has been identified for island Southeast Asia. This finds its earliest expression on Taiwan by about 3500 BC, with progressively later dates as one moves south into the Philippines (3000 BC), to Timor (2000 BC) and the Bismarck Archipelago (1500 BC). Thence, expansion proceeded into Melanesia and greater Oceania. The abrupt nature of this, in archaeological terms, may be appreciated through excavations at Gua Uattamdi on Kayoa Island. There is no evidence for occupation there before the 1500 to 1 BC, when the prehistoric period saw the arrival of people who made red-slipped pottery vessels, bone points, polished adzes, shell bracelets and beads. This assemblage is matched in the Philippines, Sulawesi, and in the well-known Melanesian Lapita sites. On Gebe island, by contrast, a hunting, gathering and fishing economy which began at least by 31,000 years BC continued, unaffected by any new arrivals until the opening of the area to the international spice trade two millennia ago (Bellwood 1998).

The languages spoken on the mainland of Southeast Asia have been divided into five phyla by Pejros and Schnirelman (1998): Austroasiatic (AA), Austronesian (AN), Tai-Kadai, Miao-Yao and Sino-Tibetan. Austronesian languages dominate in island Southeast Asia. AA languages are most relevant to any consideration of the spread of agriculturalists because of their wide distribution and proposed time depth. There are three major divisions, Munda in eastern India, Mon-Khmer and Nicobarese. The major languages are Vietnamese and Khmer. Mon was formerly widely spoken in Thailand and Burma, but has now been superceded by Sino-Tibetan and Tai languages. The general distribution suggests that a former broad band of related languages has been divided and isolated by later intrusions. There is also a consistent body of evidence, which suggests that AA languages were formerly spoken in significant areas of southern China (Hashimoto 1972, Norman and Mei 1976). Zide and Zide (1976) and Mahdi (1998) have identified a series of cognates, which link widely distributed AA languages. The latter, for example, has reconstructed the Proto AA form for husked rice from Munda, Mon-Khmer, Palaung, Viet-Muong, Old Mon, Lamet, and Vietnamese cognates.

Almost a century ago, Schmidt (1906) proposed that AA and Austronesian languages had a common ancestor, which he named Austric. This lay in abeyance until Reid (1994) was able to give detailed consideration to Nancowry, a language of the Nicobarese division of AA. He found convincing evidence in the form of morphemes for an ancient link between AA and AN, thereby strongly supporting the Austric hypothesis. Diffloth (1994) has added supportive lexical evidence. Blust (1996) adopted both when he explored the archaeological implications of an Austric macrofamily. He proposed that rice cultivation was taken by speakers of the proto Munda languages into eastern India through the Brahmaputra valley, while proto Viet, Mon and Khmer languages were brought respectively down the Red, Chao Phraya Salween, Chindwin, Irrawaddy and Mekong rivers. This general expansion to the south has been supported by Pejros (1998). He has suggested, on the grounds of linguistic palaeontology, that Proto-Austric was spoken north of the tropics, and that it began to dissolve into daughter languages in the 8th to 9th millennia BC. By the end of the 5th millennium, he suggests, AA included several branches, including Munda and Mon-Khmer. By the end of the 4th millennium, Proto Mon-Khmer had divided into Khmer, Viet-Muong and Bahnaric. Reconstructed linguistic terms for the proto languages exclude a coastal origin, and he has suggested that the AA homeland was in the Middle Yangzi Valley. By the same reasoning, Blust (1996) has placed the proto-AN homeland on Taiwan, or the adjacent mainland. He also notes, most importantly, that the reconstructed terms for domestic rice belongs to this proto AN.

These two quite separate syntheses, on the one hand archaeological and the other linguistic, may now be considered together. The key archaeological finding is that the transition to rice cultivation took place in the middle reaches of the Yangzi, and found its early expression in village communities represented by Bashidang and Pengtoushan. Pejros and Blust both place the homeland of Proto AA languages this far north, although the latter prefers the upper reaches of the river. This is not sustained by archaeology, which rather indicates an upriver expansion of agricultural communities into Sichuan. The expansion downstream is archaeologically documented at the well-known site of Hemudu, and the settlements which lie beside Lake Taihu. These are up to two millennia later than Pengtoushan, and bring rice cultivation within striking distance of the straits of Taiwan and tangible relations with early AN.

The expansion to the south is now supported by both archaeological and linguistic evidence, but the linguists' dates for the separation of AA into its component parts do not match those obtained by archaeology. The former estimates are up to two thousand years earlier than radiocarbon determinations from archaeological sites. However, at a conference held in Cambridge during 1999, this very topic of time depth in linguistic change revealed how shaky are the grounds for estimating chronologies.

The formulation of this model for the demic expansion of agriculturalists based on archaeological and linguistic evidence is only a faltering step forward in our investigations. Certainly, it has generated debate and differences of opinion (Meacham 1991). However, the development of the polymerase chain reaction and the opening of the whole field of ancient DNA provide a possible means to test and refine it. This will involve the successful extraction and sequencing of mtDNA from human remains from East and Southeast Asian archaeological sites, thus following in the pioneering footsteps of Hagelberg et al. (1999) and Melton et al. (1995) in the analysis of DNA from Pacific populations. A second approach involves the analysis of DNA from dog bones. The Southeast Asian dog first appears in the archaeological record with early agricultural communities, and only then appears in hunter-gatherer sites. Cranial morphology shows descent from the wolf, and there are no wolves in Southeast Asia. However, the Chinese wolf, *Canis lupus chanco*, is the most likely ancestor. Initial attempts to sequence canid DNA from prehistoric Southeast Asian sites have supported this derivation.

3. Studies of population structure: Assessing the effects of cultural practices such as polygyny, patrilocality, and uxorilocality on the patterns of variation on the X and Y chromosome, the autosomes, and mitochondrial DNA

Sir Edmund Leach (1973) once commented that the kinship structure of a prehistoric community lay forever sealed in a black box, beyond the visible horizon of archaeological research. This was a provocative and challenging pronouncement, which can be revisited in the light of new developments in the study of ancient DNA.

Southeast Asian prehistoric cemeteries, when opened on a sufficient scale, never fail to reveal a pattern. As Dr Rachanie (Chapter 3) will describe for Khok Phanom Di, the pattern at that site saw the formation of burial clusters which, superimposed over time, have allowed us to propose a sequence of about 17 generations (Higham and Thosarat 1994). At the Bronze Age sites of Nong Nor and Ban Lum Khao, we found that the dead were interred in rows. At Noen U-Loke, an Iron Age settlement, we again encounter clusters. These groups incorporate the remains of men, women, children and infants. The infants were often interred in mortuary jars at the feet, or beyond the head, of an adult. Men and women were regularly found in association, even lying alongside each other, or in the same grave. The Chinese archaeological record includes many similar patterned cemeteries.

At Khok Phanom Di, our reconstruction of genealogies covering about four centuries incorporated men and women interred alongside each other. But we do not know if their relationships were affinal, consanguinal or, unlikely as it seems, fortuitous. So a man might be the woman's husband, brother, father or son. Again, infants buried at the feet of an adult are probably related, but how? Answering these questions should make it possible to investigate such issues as matrilocality. In the case of Khok

Phanom Di, we have concluded that women played a pre-eminent role as makers of outstandingly beautiful pottery vessels. Is it possible, through the study of ancient DNA, to demonstrate that successive generations of women provided genetic continuity? Or in the case of an Iron Age community in which men were increasingly involved in warfare, that a group was patrilocal? On a regional scale, we find in Northeast Thailand that there was a tight regional distribution of preferred ceramic styles. Does this have a correlate in ethnicity and population structure in that each grouping was endogomous? These are all variables, which, if illuminated, would at least prise open Leach's black box.

Acknowledgements

I would like to thank Mark Seielstad and Li Jin for inviting me to contribute to this publication, and the staff of Yunnan University for their hospitality.

References

Ammeman, A., Cavalli-Sforza, L. 1984. *The Neolithic Transition and the Genetics of Populations in Europe.* Princeton: Princeton University Press.

Bellwood, P. 1997. *Prehistory of the Indo-Malaysian Archipelago.* 2nd ed. Honolulu: University of Hawaii Press.

Bellwood, P. 1998. The archaeology of Papuan and Austronesian prehistory in the Northern Moluccas, Eastern Indonesia. IN: Blench, R. and Spriggs, M. (eds.), *Archaeology and Language II. Correlating Archaeological and Linguistic Hypotheses*. London: Routledge, pp. 128-140.

Blench, R.M., Spriggs, M. 1998. General introduction. IN: Blench, R. and Spriggs, M., (eds.), *Archaeology and Language II. Correlating Archaeological and Linguistic Hypotheses.* London: Routledge, pp. 1-19.

Blust, R. 1996. Beyond the Austronesian homeland: the Austric hypothesis and its implications for archaeology. IN: Goodenough, W.H. (ed.), *Prehistoric Settlement of the Pacific. Transactions of the American Philosophical Society* 86:117-140.

Bulbeck, D. 1999. Current biological anthropological research on Southeast Asia's negritos. *Spafa J.* 9(2):14-22.

Colani, M. 1927. L'age de la pierre dans la province de Hoa-Binh. *Memoires du Service Geologique de l'Indochine* XIII:1

Crawford, G.W., Chen, Shen. 1998. The origins of rice agriculture: recent progress in East Asia. *Antiquity* 72:858-866.

Diffloth, G. 1994. The lexical evidence for Austric, so far. *Oceanic Linguistics* 33:309-322.

Green, R.C., Pawley, A. In press. Early Oceanic architectural forms and settlement patterns: Linguistic, archaeological and ethnological perspectives. IN: Blench, R. and Spriggs, M. (eds.), *Archaeology and Language III: Artifacts, Languages and Texts: Building Connections*. London: Routledge.

Hagelberg, E. 2000. Personal communication.

Hagelberg, E., Kayser, M., Nagy, M., Roewer, L., Zimdahl, H., Krawczak, M., Lió, P. et al. 1999. Molecular genetic evidence for the human settlement of the Pacific: Analysis of mitochondrial DNA, HLA and Y chromosome markers. *Phil. Trans. Roy. Soc. Lond.* B 354:141-152.

Hashimoto, O.Y. 1972. *Studies in Yue dialects 1: Phonology of Cantonese.* Cambridge: Cambridge University Press.

Higham, C.F.W., Thosarat, R. 1994. Khok Phanom Di. Prehistoric Adaptation to the World's Richest Habitat. Dallas, Fort Worth: Harcourt-Brace.

Higham, C.F.W., Lu, T. L.-D. 1998. The origins and dispersal of rice cultivation. *Antiquity* 72:867-877.

Leach, E.R. 1973. Concluding address. IN: Renfrew, C. (ed.), *The Explanation of Culture Change: Models in Prehistory*. London: Duckworth, pp. 761-771.

Mahdi, W. 1998. Linguistic data on transmission of Southeast Asian cultigens to India and Sri Lanka. IN: Blench, R. and Spriggs, M. (eds.), *Archaeology and Language II. Correlating Archaeological and Linguistic Hypotheses*. London: Routledge, pp. 390-415.

Meacham, W. 1991. Further consideration of the hypothesized Austronesian Neolithic migration from South Chinato Taiwan and Luzon. *Bull. Indo-Pacific Prehistory Association* 11:398-407.

Melton, T., Peterson, R., Redd, A. J., Saha, N., Sofro, A. S. M., Martinson, J., Stoneking, M. 1995. Polynesian genetic affinities with Southeast Asian populations as identified by mtDNA analysis. *Am. J. Hum. Genet.* 57:403-414.

Norman, J., Mei, T. 1976. The Austroasiatics in ancient South China; some lexical evidence. *Monumenta Serica* 32:274-301.

Pei, Anping. 1998. Notes on new advancements and revelations in the agricultural archaeology of early rice domestication in the Dongting Lake region. *Antiquity* 72:878-885.

Pejros, I., Shnirelman, V. 1998. Rice in Southeast Asia: A regional interdisciplinary approach. IN:Blench, R. and Spriggs, M. (eds.), *Archaeology and Language II. Correlating Archaeological and Linguistic Hypotheses*. London: Routledge, pp. 379-389.

Reid, L.A. 1994. Morphological evidence for Austric. *Oceanic Linguistics* 33:323-344.

Renfrew, C. 1987. Archaeology and Language: the Puzzle of Indo-European Origins. London: Jonathon Cape.

Schmidt, W. 1906. Die Mon-Khmer Volker: ein Bindeglied Zwischen Volkern Zentralasiens und Austronesiens. Braunschweig.

Spriggs, M. 1998. From Taiwan to the Tuamotus: absolute dating of Austronesian language spread and major sub groups. IN: Blench, R. and Spriggs, M. (eds.), *Archaeology and Language II. Correlating Archaeological and Linguistic Hypotheses*. London: Routledge, pp. 115-127.

Van Driem, G. 1998. Neolithic correlates of ancient Tibeto-Burman migrations. IN: Blench, R. and Spriggs, M. (eds.), *Archaeology and Language II. Correlating Archaeological and Linguistic Hypotheses*. London: Routledge, pp. 67-102.

YPM (Yunnan Provincial Museum). 1981. The Baiyangcun site at Binchuan County, Yunnan Province (in Chinese). *Kaogu Xuebao* 1981:349-68.

Yuan, Jairong, Zhang, Chi. 1999. The origins of pottery and rice cultivation in China. Newsletter of the Grant-in-Aid Program for COE Research Foundation of the Ministry of Education, Science, Sports and Culture in Japan 2(1):3-4.

Zhao, Zhijun.1998. The middle Yangtze region in China is one place where rice was domesticated: phytolith evidence from Diaotonghuan cave, Northern Jiangxi. *Antiquity* 72:885-897.

Zide, A.R.K., Zide, N.H. 1976. Proto-Munda cultural vocabulary: evidence for early agriculture. IN: Jenner, P.N., Thompson, L.C. and Starosta, S. (eds.), *Austroasiatic Studies II*. Honolulu: Oceanic Linguistics Special Publication 13, pp. 1205-1334.

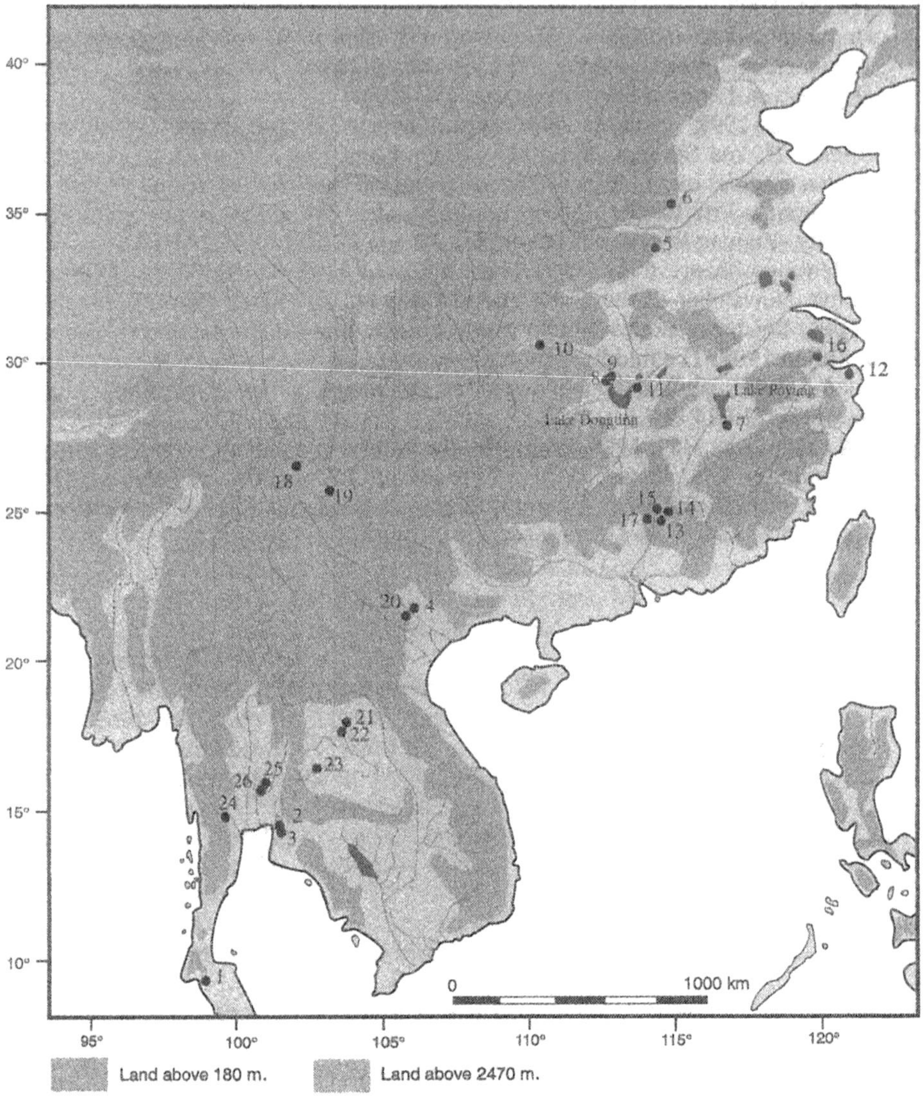

The main archaeological sites mentioned in the text: **1.** Lang Rongrien, **2.** Khok Phanom Di, **3.** Nong Nor, **4.** Son Vi, **5.** Peiligang, **6.** Cishan, **7.** Xianrendong and Diaotonghuan, **8.** Pengtoushan, **9.** Bashidang, **10.** Chengbeixi, **11.** Fenshanbao, **12.** Hemudu, **13.** Shixia, **14.** Xincun, **15.** Chuangbanling and Niling, **16.** Liangzhu, **17.** Nianyuzhuan culture sites, **18.** Baiyangcun, **19.** Dadunzi, **20.** Phung Nguyen, **21.** Ban Chiang, **22.** Non Kao Noi, **23.** Ban Lum Khao, **24.** Ban Kao, **25.** Non Pa Wai, **26.** Ban Tha Kae.

Figure 1. Map.

HUMAN DIVERSITY AND LANGUAGE DIVERSITY

WILLIAM S.-Y. WANG

Department of Electronic Engineering, City University of Hong Kong

Since language is the defining trait of our species, human evolution and linguistic evolution are obviously closely intertwined. Recent studies in genetics suggest that anatomically modern humans emerged at a very late date, perhaps 50 kys (Bertrapetit 2000; Thompson 2000). This dating is consistent with the onset of an unprecedented degree of cultural innovations, in both quality and quantity, as revealed in the archaeological record (Klein 1999). We share the belief with many students of human prehistory that the evolution of anatomically modern humans, the emergence of language, and the burst of cultural innovations, including extensive cave art and sailing across broad expanses of water are events which are all closely linked to each other.

Culturally, there have been several major transitions separating us from our prehistoric ancestors—such as the use of fire, the invention of tools, the advent of agriculture, etc. Similarly, there must have been major transitions which led from the primitive growls and howls of our ancestors to the intricate languages we have today. We cannot recover language evolution in the very distant past in ways comparable to those of the archaeologist, since the earliest 'material remains' of language, i.e., ancient texts, date back no farther than several millennia. However, linguists have developed methods of reconstruction and taxonomy which are helpful toward an interdisciplinary understanding of the diversity of peoples.

Indeed the identity of a people is often intimately coupled to the language it speaks. Linguistic grouping has been taken, time and again, to be the first criterion for sorting out human diversity. The celebrated diagram published by Cavalli-Sforza et al. (1988), comparing a genetic tree with a linguistic tree, was an eloquent statement on the important parallelisms between genetic evolution and linguistic evolution on a global scale. More locally, when a method developed to quantify genetic affinity was applied to a chain of languages in Micronesia, it was found to yield comparable results (Cavalli-Sforza and Wang 1986, reprinted in Wang 1991).

At the same time, however, languages and genes do go their separate ways, and such cases are not hard to find. When one ethnic group conquers another ethnic group, the common language eventually arrived at may be that of the conqueror, or that of the conquered. The latter is clearly the case with the Manchus, an Altaic people from northeastern China who founded the Qing dynasty and ruled the entirety of China for nearly 300 years. Although there are numerous monuments and documents which attest to the glory of their long reign, the Manchu language has been all but replaced by the language of the Han majority. Li (2000, p. 15) describes the situation this way.

> *"A survey done in the People's Republic of China in the 1950's found that quite a few elderly Manchus who lived in the more remote regions of Manchuria could still speak Manchu. Those over thirty years old were likely to understand it, while the younger generation could neither speak or [sic] understand it. Since then, anthropologists and linguists doing research in northern Manchuria have been reporting on a rapidly dwindling number of Manchu speakers. By the 1990s Manchu speakers have become nearly non-existent."*

Such cases of language displacement, by no means rare, remind us that genes and languages can and do go separate ways. While they match in the default case, we should not be disturbed when their phylogenies do not agree. In fact, the cases of mismatch are in a sense more interesting since they may reveal displacement events long ago which would be difficult to uncover otherwise.

Potential contributions from linguistics on the question of human diversity come under three headings:

1. To establish genetic groups and subgroups of languages.
2. To locate the homeland of speakers of ancient languages.
3. To date splits among languages.

The study of language prehistory has a distinguished tradition in many cultures. In China, reconstructing the rhymes of ancient poetry reached a high level of scholarship in the 16th century. In the West, historical linguistics traces its roots to a famous lecture given in 1786 by William Jones. The following paragraph with which he announced the genetic relatedness among some of the languages in Europe and in Asia is perhaps the most often quoted in linguistics:

> *"The Sanskrit language, whatever be its antiquity, is of a wonderful structure; more perfect than the Greek, more copious than the Latin, and more exquisitely refined than either, yet bearing to both of them a strong affinity, both in the roots of verbs and in the forms of grammar, than could possibly have been produced by accident; so strong indeed, that no philologer could examine them all three, without believing them to have sprung from some common source, which perhaps no longer exists; there is a similar reason, though not quite so forcible, for supposing that both the Gothick and the Celtick, though blended with a very different idiom, had the same origin with the Sanskrit; and the old Persian might be added to the same family, if this were the place for discussing any questions concerning the antiquities of Persia." (Quoted in Cannon 1991:31)*

Building upon Jones's insight, a great deal has been achieved toward clarifying the relationships among the 6000 or so languages spoken in the world today. The reconstruction of the Proto-Indo-European, the "common source" that Jones conjectured in the above paragraph, together with the light it sheds on civilizations of some 7,000 years ago, has become a standard in scholarship to be emulated everywhere. Many proto-languages of similar time depths have been reconstructed.

Currently, there is a spectrum of positions on how much time depth is recoverable in language for determining genetic relationships. At one end of the spectrum, some linguists have been reluctant to venture beyond the time depth established by Indo-European studies. Since a living language is constantly changing, these linguists believe that nothing reliable will be left of the original language after 7,000 years to be of diagnostic value. Although this ceiling of 7,000 years has never been objectively justified, it seems to reflect a bias from Indo-European studies. At the other end of the spectrum, some linguists propose global etymologies, roots of words which can be found in all major phyla. These linguists believe that all the world's languages can be traced to a single monogenetic source.

While monogenesis is the dominant view today, probabilistic considerations actually favor a scenario in which language was invented independently at many sources, i.e., polygenesis (Freedman and Wang 1996). In pondering these issues, we should also take into account the effects of global events such as major glaciations, which must have scrambled human populations extensively by forcing distant migrations. It would be difficult to establish linguistic lineages across such barriers of panmixia.

Although methods of taxonomy are not nearly as well developed in linguistics as in biology, nonetheless a general picture is emerging, largely thanks to the pioneering efforts of Joseph H. Greenberg of Stanford University. Figure 1 shows the dozen or so phyla he proposes for the languages of the world. This classification is discussed by Ruhlen (1991). While most of the details remain to be worked out, his proposal is the first major framework within which future research can be anchored. The phylum that Greenberg has been investigating in depth himself is one he calls Eurasiatic. As shown in Figure 2, the Eurasiatic phylum has Indo-European as one of its branches, but also comprises many other branches as well, including the enigmatic Ainu language, which has been considered by most to be a linguistic isolate. Greenberg's results (2000), which have been just published, are sure to elicit very different responses from linguists of various persuasions.

Quite independent of Greenberg's research, a group of Russian linguists, led by the late Illich-Svitych, have also proposed a large phylum of languages, which they call Nostratic. For some discussion of the Nostratic proposal, see the anthrology edited by Salmons and Joseph (1998). It is instructive to compare the memberships of the two proposals, as seen in Table 1. Much of the original work on the two proposals was done during the decades when communication across the continents was hampered by political curtains, and the sharing of data was difficult. Recent years have seen closer interactions between the linguists of the U.S. and Russian, with the encouraging result of increasing convergence in their views.

Another phylum of great interest is Dene-Caucasian. The proposal by Sergei Starostin (1990), a linguist at the Moscow State University, is shown in Figure 3. Again, while some members of the phylum may be firmly established, such as Sino-Tibetan, much work needs to be done for the proposal to reach general acceptance. An example of recent progress here is the finding of Ruhlen (1998), on the Yeniseian and Na-Dene, which are two branches of the Dene-Caucasian. This finding of 36 common etymologies is of special interest since it definitively connects languages which are currently distributed on opposite sides of the Pacific.

There is still no consensus regarding the distant affiliations of the Chinese language. This is reflected in a monograph edited by Wang (1995), in which E.G.Pulleyblank discusses the connection between the Chinese and Indo-European. Laurent Sagart (see Wang 1995) discussed the Chinese and Austronesian. In the same monograph, Starostin shows the number of basic words, defined by Sergei Yakhontov (see Wang 1995), shared among these language groups. In Table 2, Starostin's numbers have been converted to

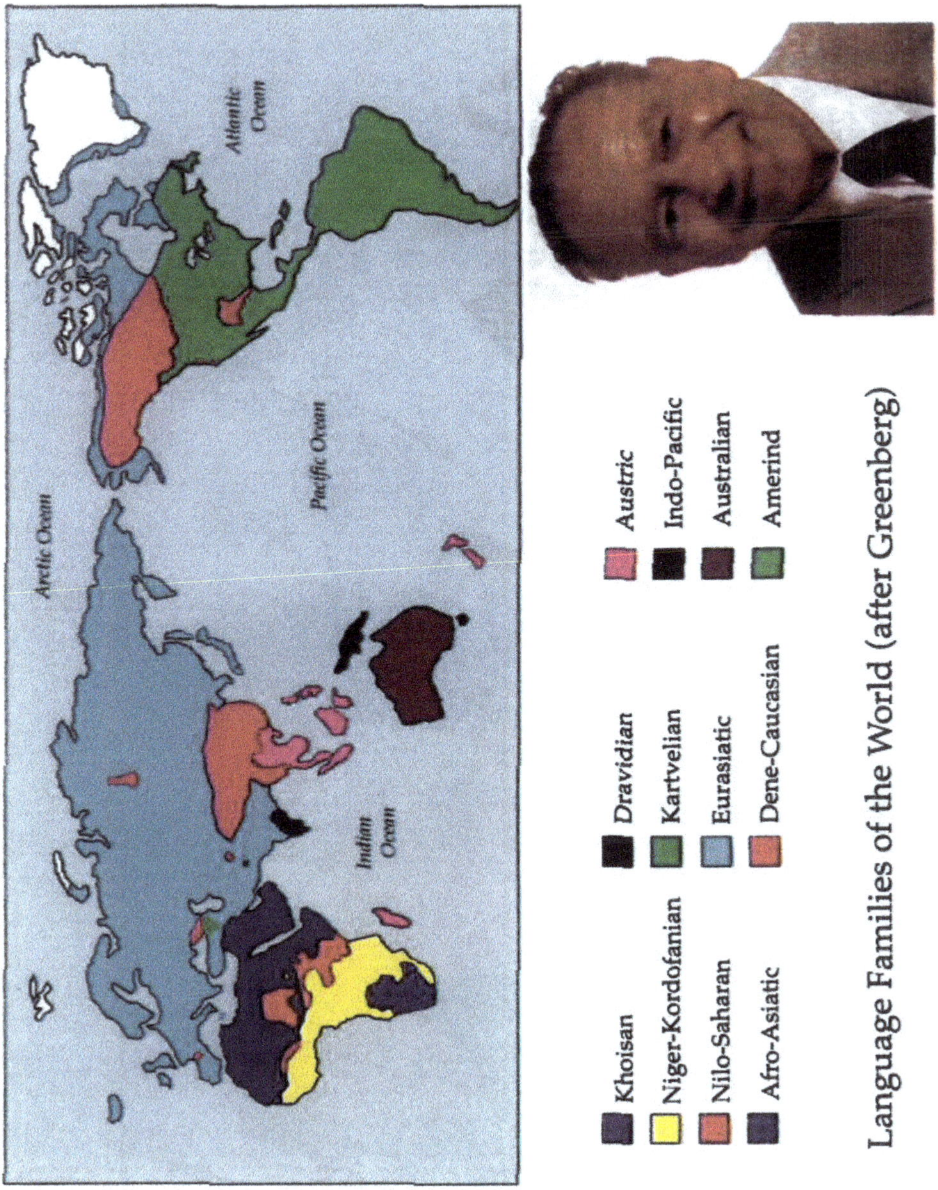

Figure 1. The language phyla of the world, proposed by Joseph H. Greenberg.

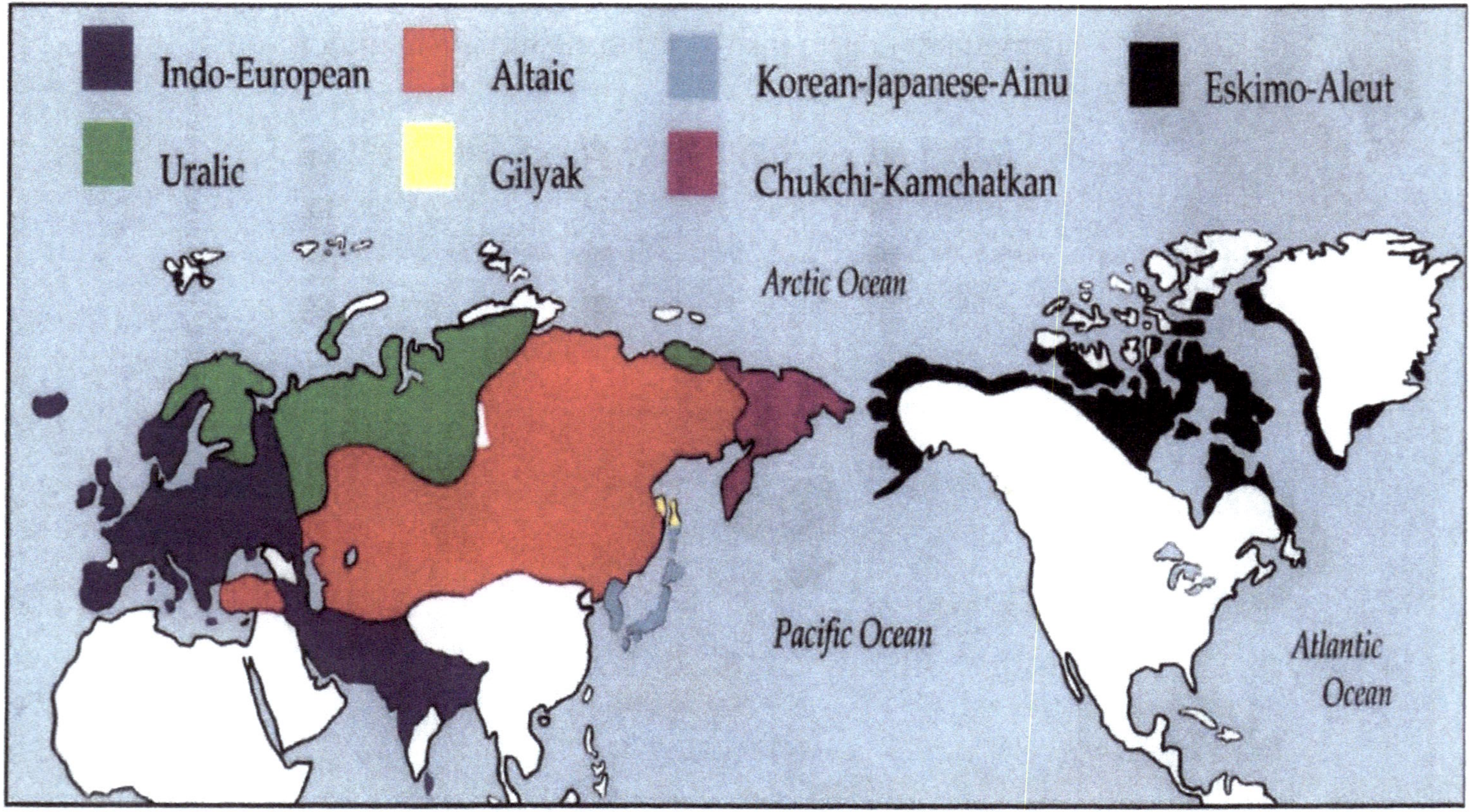

The Eurasiatic Family

Figure 2. The Eurasiatic phylum of languages, proposed by Joseph H. Greenberg.

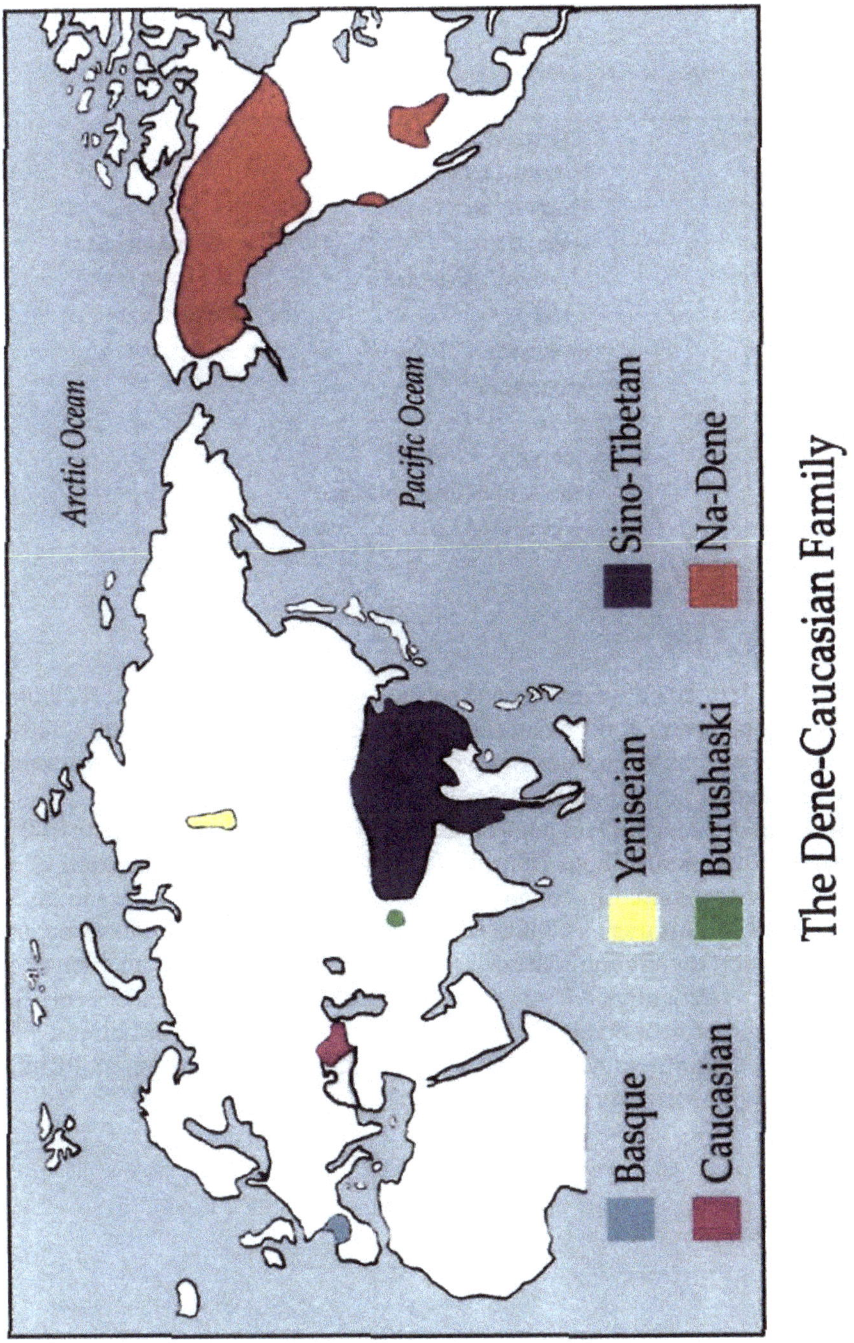

Figure 3. Map of Dene-Caucasian phylum of languages, proposed by Sergei Starostin.

Table 1. Comparison between two classifications, Nostratic and Eurasiatic

Nostratic [1]	Afro-Asiatic	
∀	Elamo-Dravidian	
∀	Kartvelian	
∀	Indo-Hittite	Eurasiatic[2]
∀	Uralic-Yukaghir	∀
∀	Altaic	∀
∀	Korean	∀
	Japanese	∀
	Ainu	∀
	Gilyak	∀
	Chukchi-Kamchatkan	∀
	Eskimo-Aleut	∀

[1] Illich-Svitych, 1971-1984
[2] Greenberg, 2000

percentages. It can be seen there that the subset of the Chinese, Tibeto-Burman, Caucasian and Yeniseian does show a significantly closer relationship internally than any member has with either the Indo-European or Austronesian.

The Dene-Caucasian languages are largely found in the north; the major exception being some Tibeto-Burman languages which have migrated deep into Southeast Asia. In complementary distribution to the linguistic developments in northern Asia, the languages of southern Asia have become grouped under the phylum Austric. The reality of this phylum has been considerably strengthened in recent years with the discovery of morphological correspondences by Reid (1994). The Austric phylum is a far-flung group, comprising well over 1000 languages. According to Ruhlen (1991), the major subgroups are as follows:

Austric:

I. Miao-Yao
II. Austro-Asiatic
 a. Munda
 b. Mon-Khmer. e.g. Wa, Vietnamese.
III. Austro-Tai
 a. Daic. e.g. Zhuang, Thai, Lao.
 b. Austronesian.
  i. Eastern = Oceanic, e.g. Hawaiian.
  ii. Western. e.g. Malagasy, Tagalog.

A leading authority on Austric languages is Robert Blust (1996) of the University of Hawaii. Although Blust's latest classification of Austric may differ somewhat from that of Ruhlen, he offers the following approximate dates of divergence, which provide a useful temporal framework.

Proto-Austric	8,500 BP
Proto-Austronesian	6,500
Proto-Oceanic	4,000

Reviewing the archaeological evidence, Blust suggests that the last unity of the Austric phylum may have been at the Yunnan-Burma border, splitting into various families, which then spread into South China, Southeast Asia. The paths these early migrants took probably followed the courses of the great rivers of Asia.

Table 2. The relation of Chinese to other groups of languages, shown as the percentage of apparent cognates from 35-word list of Yakhontov

	OC	PTB	PNC	PY	PIE
Old Chinese					
Proto-Tibeto-Burman	74				
Proto-North-Caucasian	43	51			
Proto-Yenisseian	34	40	57		
Proto-Indo-European	23	14	17	17	
Proto-Austronesian	14	11	11	11	14

We are far from having a conclusive prehistory of Asia, though scholars are beginning to bring together evidence from archaeology, genetics and linguistics. If we accept the three language phyla discussed above, then a plausible scenario from linguistics is this. Early humans entered East and

Southeast Asia, bringing with them two linguistic phyla. the Dene-Caucasian in the north and the Austric in the south. Their domains were later supplanted by the Eurasiatic phylum, particularly the Altaic family and the Indo-Iranian branch of the Indoeuropean family.

The Altaic family of languages stretches like a belt across Central Asia, stretching from Turkey in the west and extending to the Pacific in the east over several millennia. Only in recent centuries did Russian, a member of the Slavic branch of the Indo-European family, colonize large regions of northern Asia. The Indo-Iranian languages have moved into West Asia and South Asia, where they claim large communities of speakers in Iran, India and Pakistan. The expansion eastward of the Eurasiatic phylum covers over much of the territory earlier occupied by speakers of the Dene-Caucasian and Austric. With a few notable exceptions, such as Chinese, the earlier languages have been consistently shrinking as the Eurasiatic languages gained the upper hand.

Much of the evidence linguists offer is based on vocabulary. In any language, the vocabulary contains words which are more cultural, such as: tennis, television, tea, etc. Cultural words are frequently adopted from language to language, and hence are not stable indicators of genetic relations. On the other hand, all languages also have basic words which are much more stable, such as: water, hand, and tree. Although basic words do get adopted, they are relatively stable. As Morris Swadesh (1952) proposed in the 1950s, they provide a source of quantitative data for studying relations among languages.

Table 3 presents in tabular form one of the lists of 100 basic words Swadesh (1952) proposed that has gained wide acceptance in linguistic research. Various criticisms have been voiced against the concept of basic words in general, and against this list of 100 words in particular. Some scholars feel that the list is too inclusive, and whittle it down to fewer words. The table Starostin constructed, upon which Table 2 is based, uses a list of 35 words proposed by Yakhontov. In Table 3, these 35 words are shown in italics. As can be seen in the table, 32 of the 35 are in the Swadesh list. The three words Yakhontov proposes not in the Swadesh list are: salt, wind, and year.

Basic words as a method in studying linguistic prehistory has been used primarily in two contexts. One is to show degrees of affinity, as Starostin (1990) does in Table 2. The other is to estimate dates of the linguistic split. A central problem in the historical study of language is that of sorting out linguistic traits which are vertically transmitted as opposed to those which are horizontally transmitted. The former mode is also called inheritance, and the latter mode is also called borrowing. The problem is extremely difficult because any linguistic trait can be transmitted either vertically or horizontally.

Figure 4 illustrates one approach to this problem in the form of a family tree for the Austronesian languages of Taiwan. Using standard methods of cluster analysis, I constructed a tree on the basis of a table of numbers of shared words among these languages (Wang 1989). Such trees are of course

Table 3. List of 100 basic words, proposed by Morris Swadesh. A smaller subset of 32 words - plus *salt*, *wind*, and *year* - proposed by Sergei Yakhontov are shown in italics

	Nature	Body	Animal	Verb	Adjective	Misc
1	ashes	belly	bird	bite	all	earth
2	bark	*blood*	claw	burn	big	*I*
3	cloud	*bone*	*dog*	come	black	*name*
4	*fire*	breast	feather	*die*	cold	night
5	leaf	*ear*	*fish*	drink	dry	not
6	man	*egg*	*horn*	eat	fat	*one*
7	*moon*	*eye*	*louse*	fly	*full*	road
8	mountain	foot	*tail*	*give*	good	
9	person	hair		hear	green	that
10	rain	*hand*		kill	long	*this*
11	root	head		*know*	many	*thou*
12	sand	heart		lie	*new*	*two*
13	seed	knee		say	red	we
14	smoke	liver		see	round	*what*
15	star	meat		sit	small	*who*
16	*stone*	mouth		sleep	warm	
17	*sun*	neck		stand	white	
18	tree	*nose*		swim	yellow	
19	*water*	skin		walk		
20	woman	*tongue*				
21		*tooth*				
	salt					*year*
	wind					

time-honored ways of graphing vertical transmission. On the basis of the resulting tree, I was able to make another table of the *presumed* number of shared words among these languages. Comparing these two tables enabled me to detect regions of mismatch, which I interpret to be due to horizontal transmission. These horizontal transmissions are indicated on the tree by broken lines. While such a modified tree does capture both modes of transmission, the method of its construction appears to give dominance to vertical transmission.

Using similar methods, I made an attempt to estimate the date of the split of the Sino-Tibetan family of languages, as shown in Figure 5. Details of this exercise are discussed more fully in (Wang 1998). I first constructed a tree of the major dialects of Chinese, which is shown at the top of the figure. The tree shown in the middle of the figure is one I constructed for Indo-European, following identical procedures. The encouraging result when comparing the two trees is that the 'height' of the tree for Chinese dialects is approximately the same as that for the three Germanic languages in the Indo-European tree. Based on these rough yardsticks, it would seem that the Sino-Tibetan tree at the bottom of the figure should be somewhat younger than the Indo-European tree. This means that if we assume that the Indo-European tree is 7,000 years old, then the Sino-Tibetan tree would be 6,000 years old.

Although definitive support for this date of 6,000 years, arrived at from linguistic data, is hard to come by from other disciplines, there is a map drawn by the Harvard archaeologist K. C. Chang (1986) which is very suggestive. This map, shown here as Figure 6, illustrates the period of 6,000 years ago in China when for the first time there was wholesale interaction among the many cultural spheres, based on archaeological finds. The melting together of these many cultures led Chang to refer to the period as 'initial China'. There is, then, an encouraging convergence of results here between archaeology and linguistics.

With the dramatic advances made by genetics in recent years, there is accumulating an ever increasing body of genetic data that can be compared with archaeological and linguistic hypotheses. Such comparisons will surely deepen our understanding of the nature of human diversity and linguistic diversity, whether or not genetic and linguistic maps always agree. In either case, it is certain that we had only one past, and mismatches between the maps can yield important insights on when genes and languages went separate ways.

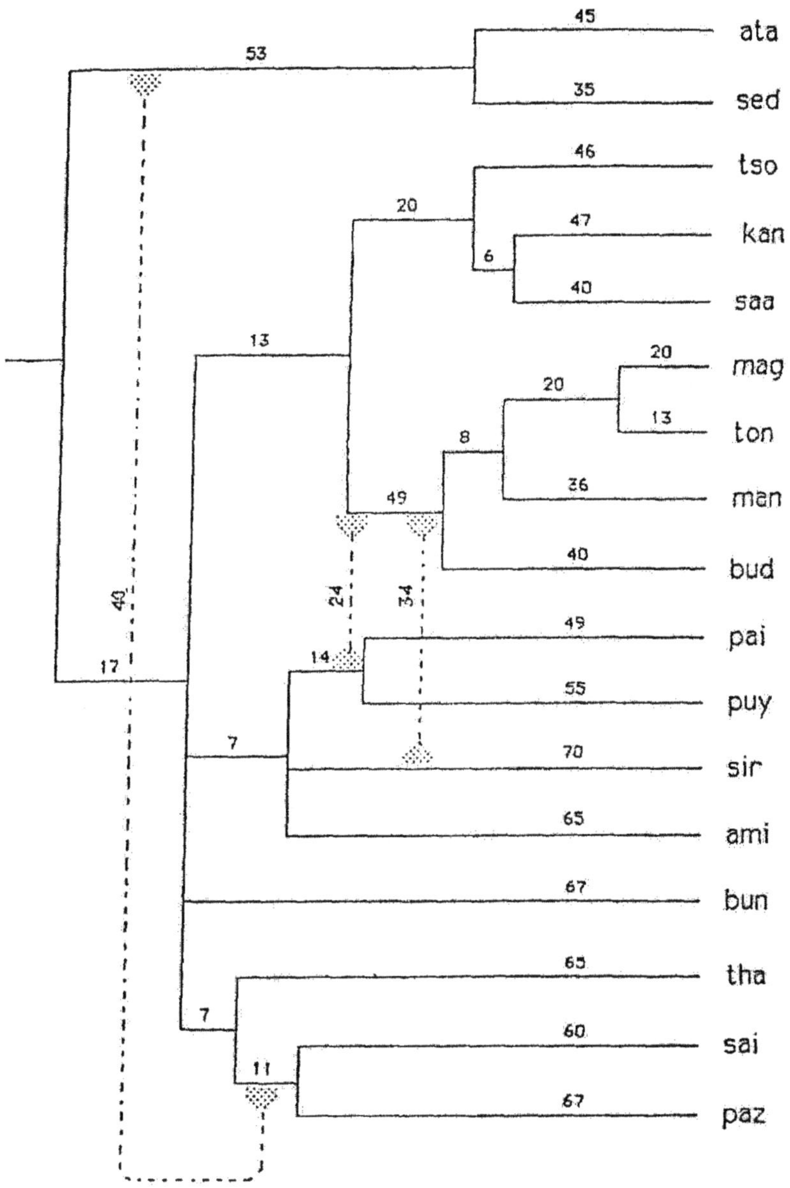

Figure 4. Vertical and horizontal transmission in the languages of Taiwan.

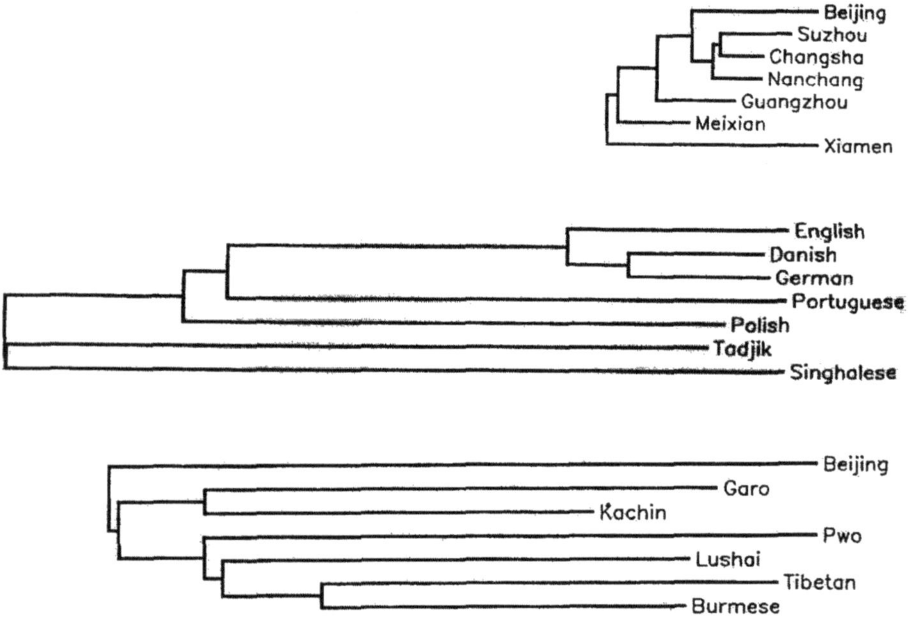

Figure 5. Additive trees of Chinese, Indo-European, and Sino-Tibetan.

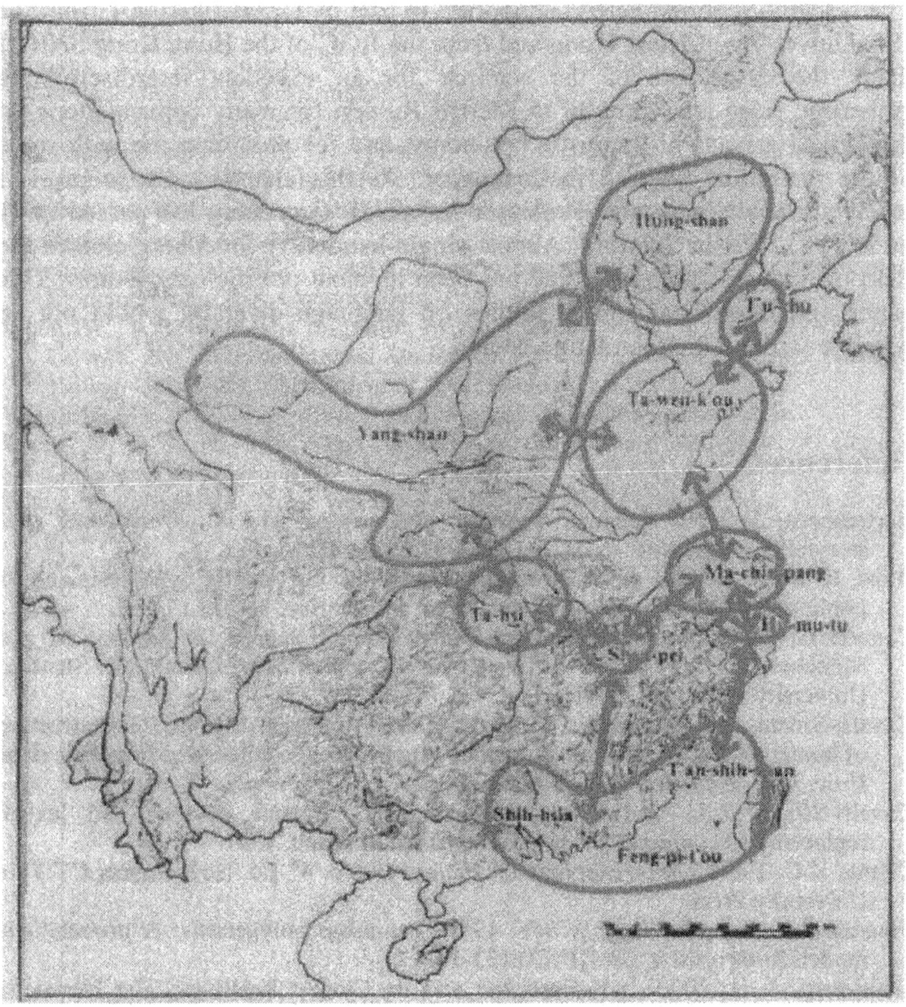

Figure 6. China in prehistory as revealed by archaeology. [Figure adapted from K.C. Chang, p. 235.]

Acknowledgements

The research reported here is supported in part by Grant #9010001 from the City University of Hong Kong and from the RGC of the Hong Kong SAR. I thank the organizers of the seminar for an excellent interdisciplinary gathering. I am also grateful to Merritt Ruhlen for many conversations on theoretical aspects of linguistic taxonomy, and for providing me with some of the materials included in this paper. As this chapter goes to press, I received the sad news that Professor Joseph H. Greenberg has passed away on May 7, 2001 in Stanford. Almost single-handedly, Greenberg created the field of linguistic taxonomy and has been its most prolific contributor. This paper and numerous similar studies on language diversity would not be possible without the foundation he laid.

References

Bertranpetit, J. 2000. Genome, diversity, and origins: The Y chromosome as a storyteller. *Proc. Natl. Acad. Sci. USA* 97:6927-6929.

Blust, R. 1996. Beyond the Austronesian homeland: the Austric hypothesis and its implications for archeology. *Trans. Amer. Philos. Soc.* 86(5):117-160.

Cannon, G. 1991. Jones's Sprung from Some Common Source. IN: Lamb, S.M. and Mithchell, E.D. (eds.), *Sprung from Some Common Source*. Stanford: Stanford University Press, pp. 23-47.

Cavalli-Sforza, L.L., Piazza, A. , Menozzi, P. and Mountain, J. 1988. Reconstruction of human evolution: Bringing together genetic, archeological and linguistic data. *Proc. Natl Acad. Sci. USA* 85:6002-6006.

Cavalli-Sforza, L.L. and Wang, W.S-Y. 1986. Spatial distance and lexical replacement. *Language* 62:38-55. Reprinted in Wang, 1991.

Chang, K.C. 1986. *The Archeology of Ancient China.* 4th Ed. New Haven, CT:Yale University Press.

Freedman, D.A. and Wang, W.S-Y. 1996. Language polygenesis: A probabilistic model. *Anthropolog. Sci.* 104(2):131-138.

Greenberg, J. H. 2000. Indo-European and its Closest Relatives: The Eurasiatic Language Family. Stanford: Stanford University Press.

Klein, R. G. 1999. *The Human Career.* 2nd ed. Chicago: University of Chicago Press.

Li, G. R. 2000. *Manchu: A Textbook for Reading Documents*. Honolulu: University of Hawaii Press.

Reid, L. A. 1994. Morphological evidence for Austric. *Oceanic Linguistics* 33:323-344.

Ruhlen, M. 1991. *A Guide to the World's Languages*. Stanford: Stanford University Press .

Ruhlen, M. 1998. The origin of the Na-Dene. *Proc. Natl. Acad. Sci. USA* 95:13994-13996.

Salmons, J.C. and Joseph, B.D. (eds.) 1998. *Nostratic: Sifting the Evidence*. Philadelphia: John Benjamins Publishing Co.

Starostin, S. 1990. *A statistical evaluation of the time-depth and subgrouping of the Nostratic macrofamily. Symposium on Molecules to Culture*. Cold Spring Harbor, NY: Cold Spring Harbor Laboratory Press, p. 33.

Swadesh, M. 1952. Lexicostatistic dating of prehistoric ethnic contacts. *Proc. Am. Philos. Soc.* 96:452-463.

Thompson, R. et al. 2000. Recent common ancestry of human Y chromosome: Evidence from DNA sequence data. *Proc. Nat. Acad. Sci. USA* 97:7360-7365.

Wang, W. S-Y. 1989. The migration of the Chinese people and the settlement of Taiwan. IN: *Anthropological Studies of the Taiwan Area*. Taiwan: National Taiwan University, Department of Anthropology, pp. 15-36.

Wang, W. S-Y. 1991. *Explorations in Language*. Taiwan: Pyramid Press.

Wang, W. S-Y., ed. 1995. The Ancestry of the Chinese Language. *J. Chinese Linguistics Monograph* 8.

Wang, W. S-Y. 1998. Three windows on the past. IN: Mair, V. (ed.), *The Bronze Age and Early Iron Age Peoples of Eastern Central Asia*. Philadelphia: University of Pennsylvania Museum Publications, pp. 508-534.

BEFORE THE NEOLITHIC: HUNTER-GATHERER SOCIETIES IN CENTRAL THAILAND

RACHANIE THOSARAT

Fine Arts Department
9th Regional Office of Archaeology and National Museum, Thailand

The long term effects of a transition to agriculture have attracted much attention in archaeological circles, following the publication of Renfrew's controversial book, *Archaeology and Language* in 1987. His thesis – which was first advanced by Ammerman and Cavallli-Sforza (1984) – was that agriculture, involving as it does sedentary communities, is an economy with a built in propensity for population growth. A growing population will usually involve the foundation of new settlements, and thus an expansion of the area occupied by agricultural communities. Since people take with them their language and genes, it should be possible to trace this process through archaeology, language and human biology. It has long been known that early farmers expanded into Europe from the Balkans by exploiting the loose soils for the cultivation of wheat and barley, and the raising of domestic stock. This process, Renfrew argues, most readily explains the present distribution of Indo-European languages. It has been tested through the analysis of blood groups, which appear to be supportive.

One of the most important advances in our understanding of East Asian prehistory over the past two decades, has been the recognition of a transition from hunting and gathering to rice agriculture in the middle Yangzi Valley. The long term implications of this sequence involve a testing of Renfrew's model in only the second major region in the Old World that experienced an indigenous Neolithic Revolution, and this has been considered by Professor Higham. This paper isolates and attempts to define the hunter-gatherer societies adapted to subtropical Southeast Asia, societies which must be understood biologically in any attempt to trace and explain a proposed expansion by rice farmers originating, ultimately, in the mid Yangzi region.

The status of hunter-gatherer cultures in Southeast Asia has been distorted for two reasons. Inland groups are known, almost without exception, on the basis of occupation remains in small rock shelters. This is undoubtedly a biased sample, resulting from the favored conditions for the survival of archaeological remains in the protected environment of a cave. Occupation sites in other inland habitats are regionally rare or absent. The second problem results from the rapid rise and subsequent fluctuations in the sea level during the Holocene Period. This resulted in a huge loss of land and then an unstable shoreline when the sea level rose higher than at present. The Southeast Asian coast, and particularly the estuary, comprises one of the richest known habitats in terms of natural productivity. There is no reasonable doubt that land now under water was formerly occupied by hunter-gatherers. When raised beaches were formed after about 4000 BC, we find many prehistoric occupation sites. In Vietnam and Southern China, these are often described as Neolithic, because of the polished stone tools and pottery vessels found during excavations. Neolithic traditionally also involves food production. Yet the evidence for subsistence other than hunting, gathering and fishing in these sites is rare if not absent. The complexity of the Jomon societies in Japan reveals how a complex material culture can develop in the context of coastal hunters and gatherers.

Between 1984 and 1992, Professor Higham and I co-directed a research program in the Bang Pakong Valley of Central Thailand, in order to define more accurately, the nature of hunter-gatherer coastal groups (Higham and Thosarat 1998). We found many sites, and excavated two, Nong Nor and Khok Phanom Di. Nong Nor was occupied for a brief period measured in months rather than years, in the vicinity of 2450 BC. It was identified during the course of an intensive site survey in 1990, and excavated over three seasons in the ensuing years. By test pitting the margins of the site, it was estimated that it covered about 700-1000 square meters, of which 405 square meters were uncovered.

Boyd et al. (1996) has examined the sediment sequence in the vicinity of the site, and has reconstructed an environment dominated by a series of indented marine embayments. When projected onto his reconstructed coast during the mid third millennium BC, we find that Nong Nor and many other similar sites were located on the shore of a large embayment, which allowed the occupants shelter while giving unfettered access to open sea.

Nong Nor comprises a shell midden, into which much later Bronze Age graves have been cut. The midden varies in thickness from a few centimeters on the eastern margin to over a meter in the center. The dominant shell species is *Meretrix lusoria*, a cockle adapted to sandy beaches. On detailed examination, however, the inhabitants undertook a wide range of activities other than collecting shellfish. Many varieties of sea fish were represented among the faunal remains, including the tiger and bull sharks. They also hunted marine mammals, such as the dolphin, and the large eagle ray. Few land animals are represented, however, and the few bones recovered were nearly all worked into artifacts. The shell midden contained many hearths, and ash spreads. Some of the latter may have been used for firing ceramic vessels, for pottery sherds were relatively common. Indeed, several complete pottery vessels were found over a crouched inhumation burial of a woman. Her grave items also included a pebble used for burnishing the surface of the clay before firing, to impart a sheen to the surface. Clay anvils used to shape pottery vessels were recovered, showing beyond doubt that the hunter-gatherers of Nong Nor made burnished and incised pottery vessels at the site. They also used polished stone adzes.

Despite the most intensive inquiry, no evidence for cultivation or the raising of domestic animals was found. The absence of dogs is particularly noted. A marine inlet, fringed in all likelihood by mangrove forest, was not a location in which one would expect rice to grow, due to the impact of the tides on the salinity of available water. All the evidence points to a marine adaptation, in which the dolphins, fish and shellfish were consumed. The thin occupation layer, lack of any evidence for periodic abandonment and the small size of the site all point to a small population and brief period of occupation. Perhaps, during the course of a seasonal occupation, the pottery vessels were made for cooking, storage and in at least the one case identified, for use as mortuary offerings to the dead.

Khok Phanom Di provides a stark contrast. While its lowest layers indicate a very similar material culture and subsistence base to Nong Nor, it remained in occupation for half a millennium, from 2000 to 1500 BC (Higham and Bannanurag 1990). During this period, the mound finally covered 5 hectares, and reached a depth of 12 meters above the surrounding rice fields.

These five centuries saw significant changes in the local environment, changes that can be related to human activities. There are two key advantages for the prehistorian: inhumation graves accumulated for at least four centuries, and it appears likely that the dead were interred in family groups over the ancestors. Second, bone and other organic material are exceptionally well preserved.

The mortuary sequence may be divided into seven phases. Already by the first, we find evidence that the people suffered from anemia, probably due to a hemoglobinopathy such as thalassemia (Tayles 1999). Infant mortality was very high, reaching virtually a half of all burials during phases 2 and 3. Adult bones also show growth disruption, but the men had physically active lives and strong upper body musculature. All the evidence points to a good diet based on fish and shellfish. The people also consumed rice, but this was probably obtained by trade with newly established inland communities. The dog, previously absent, now appeared in small numbers.

During the fourth mortuary phase, the sea level fell and freshwater habitats formed within reach of the site. There are many pointers to the local establishment of rice farming, in the new stone hoes, the shell reaping knives, the rice itself, and changes in the pattern of tooth wear and health. But the following phase saw the sea level rise again, and all these indicators for agriculture fell away in frequency or disappeared. Yet the inhabitants were now buried with great wealth in terms of shell jewelry and fine ceramic vessels: one woman was interred with over 120,000 beads, and she was almost certainly a master potter. Locally made pottery vessels were probably made for exchange and their makers achieved wealth and status expressed in their funerary rites.

It is proposed that this sequence represents a vital contribution to the consideration of human expansion into Southeast Asia, bringing with it agriculture, during the late third millennium BC. Nong Nor was a hunter gatherer site occupied at or just before the earliest evidence for the establishment of inland farming communities at about 2450 BC. There is no evidence of agriculture nor domestication of dogs. The dog is an important animal, because it was descended from the wolf. There are no native wolves in Southeast Asia, making China the most logical source for the domestic dog. The initial inhabitants of Khok Phanom Di were clearly of the same culture as Nong Nor, and did not have dogs. But they did receive exotic ceramics vessels by trade, and these vessels were occasionally tempered with rice chaff. The domestic dog also appeared at Khok Phanom Di fairly early in the sequence. It is proposed that the people of Khok Phanom Di encountered intrusive agriculturalists and exchanged goods and ideas with them. Perhaps there was also a gene exchange.

Preliminary analyses of ancient dog DNA shows relationships with the Chinese dog, while attempts to obtain and analyze mitochondrial DNA from the human remains of Khok Phanom Di have shown equivocal results. Nevertheless, the analysis of ancient DNA from prehistoric sites in Southeast Asia has the potential to test the model of human expansion which is thought to have originated with rice cultivation in the Yangzi Valley during the eighth millennium BC.

Acknowledgements

I would like to thank the organizers of this program for inviting me to participate, and the Fine Arts Department of Thailand for its support. My thanks are also due to the staff of Yunnan University for their hospitality and assistance.

References

Ammerman, A., Cavalli-Sforza, L. 1984. *The Neolithic Transition and the Genetics of Populations in Europe.* Princeton: Princeton University Press.

Boyd, W.E., Higham, C.F.W., Thosarat, R. 1996. The Holocene palaeogeography of the southeast margin of the Bangkok Plain, Thailand and its archaeological implications. *Asian Perspectives* 35:139-162.

Higham, C.F.W., Bannanurag, R. 1990. *The Excavation of Khok Phanom Di: Vol. I, The Excavation, Chronology, and Human Burials*. London: The Society of Antiquaries of London and Thames and Hudson.

Higham, C.F.W., Thosarat, R. eds. 1998. *The Excavation of Nong Nor, a Prehistoric Site in Central Thailand.* Oxford: Oxford Books, Oxford and University of Otago Studies in Prehistoric Anthropology, No. 18.

Renfrew, C. 1987. *Archaeology and Language: The Puzzle of the Indo-European Origins*. London: Cape.

Tayles, N.G. 1999. *The Excavation of Khok Phanom Di, a Prehistoric Site in Central Thailand.* Vol. 5: The People. London: Society of Antiquaries.

Part II: The Peopling of Southeast Asia

THE CASE FOR AN AFRICAN RATHER THAN AN ASIAN ORIGIN OF THE HUMAN Y-CHROMOSOME YAP INSERTION

PETER A. UNDERHILL
Department of Genetics, Stanford University

CHARLES C. ROSEMAN
Department of Anthropological Sciences, Stanford University

Introduction

DNA sequence variation on the haploid non-recombining portion of the Y-chromosome coupled with its smaller effective population size makes it an informative index of population history (Jobling and Tyler-Smith 1995). While the causes remain uncertain, recent results indicate that the history of human Y-chromosomes is characterized by a reduction of variation greater than other components of the genome including mtDNA (Shen et al. 2000). This phenomenon influences the association of Y chromosome diversity with geography. Since its initial description (Hammer 1994), the Y-chromosome specific DYS287 Alu insertion element, which is localized to Yq11 in the non-recombining region, has been one of the most widely surveyed human polymorphisms associated with this paternally inherited chromosome (Santos et al. 1996; Hammer et al. 1997). DYS287 is commonly referred to by the acronym YAP (for Y-chromosome Alu Polymorphism) in which allelic states are indicated either as YAP- (ancestral) and YAP+ (derived). The YAP insertion event defines a relic bifurcation within the internal architecture of the Y chromosome genealogy (Hammer and Horai 1995) and is useful in the evolutionary analysis of populations (Hammer and Zegura 1996). YAP+ chromosomes, while observed worldwide (Hammer et al.1997), display the highest frequencies in Africa (Spurdle et al. 1994), Tibet (Hammer et al 1997; Qian et al. 2000) and Japan (Hammer and Horai 1995).

It was reported that DNA sequencing revealed that the Alu element in both African and Asian YAP+ chromosomes was inserted at identical positions within the Y-chromosome (Hammer 1994). This provided evidence that the YAP+ insertion was a unique mutational event and that both African and Asian YAP+ lineages have shared common ancestry. While this pronounced transcontinental bipolar distribution of YAP+ lineages was initially considered (Hammer 1994) consistent with an African origin, an alternative geographic origin of YAP+ has been subsequently proposed. Specifically, on the basis of compound haplotypes constructed from genotyping both the YAP locus and a G to A transition at nucleotide position 4064 within the SRY gene region, (Whitfield et al. 1995) it was cautiously proposed that the YAP+ insertion event occurred in Asia (Hammer et al. 1997, Altheide and Hammer 1997).

Support for the claim of an Asian origin of YAP hypothesis was further bolstered by a nested cladistic analysis of Y-chromosome variation based on 9 diallelic polymorphisms typed in 1544 samples including YAP and SRY4064 (Hammer et al.1998). These results were interpreted as support for the an original range expansion out of Africa, but also were presented as validation of the intriguing additional scenario that a second more recent range expansion returned a subset of Asian Y-chromosomes (the YAP+ lineage) "back to Africa" without completely replacing the indigenous ancestral YAP- characterized African Y-chromosome pool. This was a new reconstruction of modern human pre-historical migration that continues to stimulate inconclusive discussion concerning the geographic origins of the YAP insertion (Bravi et al. 2000). On the basis of a parsimonious genealogy defining 116 distinctive lineages constructed from 1 triallelic and 166 biallelic polymorphisms surveyed in over 1000 samples, a global picture, of modern human Y chromosome diversity has been created (Underhill et al. 2000). Using these results, we summarize here all the various basic haplogroup components of East Asian Y-chromosome heritage and re-evaluate the claim of an Asian origin of YAP+ lineages by presenting evidence that contradicts the hypothesis of an Asian origin of the YAP Alu insertion.

Materials and Methods

The markers, PCR protocols and haplogroup frequency data used to evaluate the global and Asian patterns of Y-chromosome binary haplotype diversity have been previously described (Underhill et al. 2000). An additional C to T transition, called PN2, (Hammer et al.1997) has been included since it further resolves the haplotype relationships within the predominantly African

YAP+ haplogroup. DNA from 1062 men belonging to 21 populations was analyzed. Details concerning the geographic affiliations of these samples are given in Underhill et al.2000. The ten haplogroups of the parsimonious Y-chromosome phylogeny are identified with Roman numerals and defined as previously described (Underhill et al.2000). With the exception of the YAP that was analyzed according to published protocols (Hammer and Horai 1995), all other Y markers were genotyped by denaturing high performance liquid chromatography (DHPLC) methodology (Underhill et al.1997, Oefner and Underhill 1998).

Results and Discussion

1. The phylogeny

The topology of the phylogeny constructed from the 168 polymorphisms is presented in Figure 1. The tree is rooted using chimpanzee sequences. Group I is distinguished from Groups II-X by three reinforcing binary markers which are characterized by two transversions and a 1 bp deletion. Haplotypes within Group I represent all lineages on one side of the most basal split in the tree. Group II, also consisting overwhelmingly of African individuals, is defined by a 1 bp insertion polymorphism and distinguished from Groups III-X by the lack of M168, a C to T transition mutation that is common to all lineages within Groups III-X. The predominantly African YAP+ lineages are localized in Group III and the exclusively Asian YAP+ associated lineages are assigned to Group IV. Figure 2 shows the detailed composition of the YAP+ specific Group III and IV haplotypes as well as the relationships of Groups V and the VI-X composite cluster relative to one another and mutation M168. The geographic frequency distributions of YAP+ Groups III and IV including major subclades are presented in Figure 3. Populations

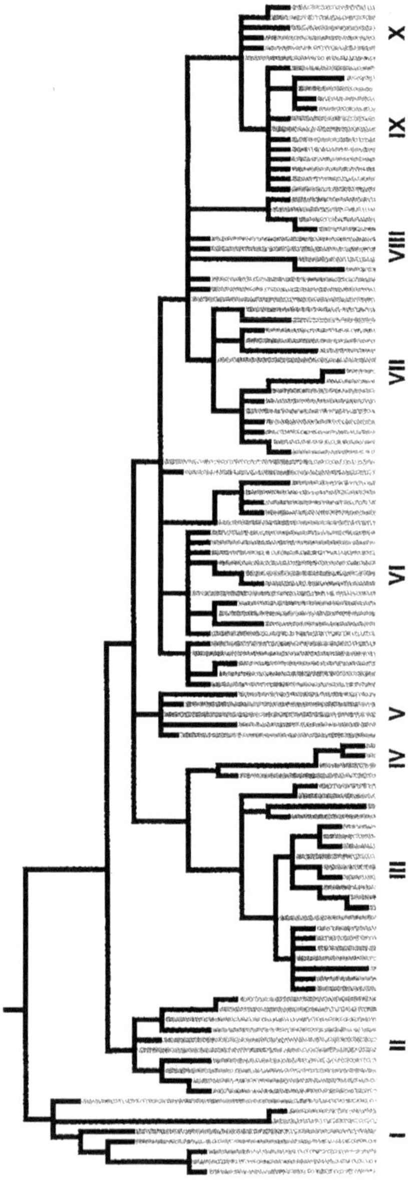

Figure 1. Topology of the maximum parsimony phylogeny of Y-chromosome binary variation. Roman numerals indicate haplogroups. Black lines are proportional to the number of mutations within each branch of the tree. Gray vertical lines indicate which haplotypes were observed.

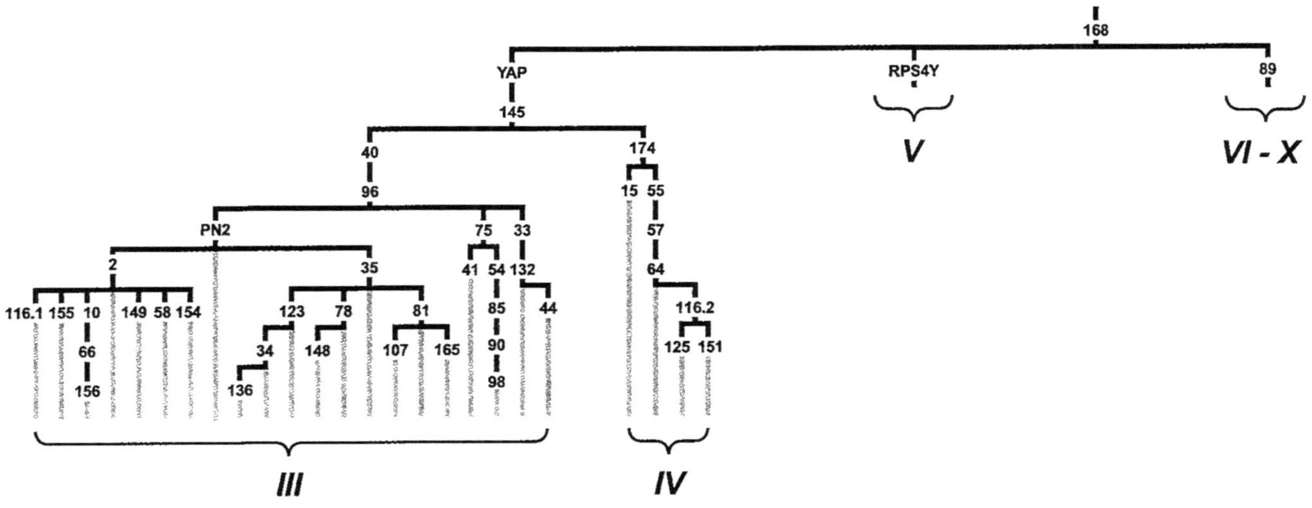

Figure 2. Details of the Group III and IV portion of the Y-chromosome phylogeny. Numbers or letters designate polymorphisms. The numbers 116.1 and M116.2 indicate the two different derived alleles for the tri-allelic marker M116.

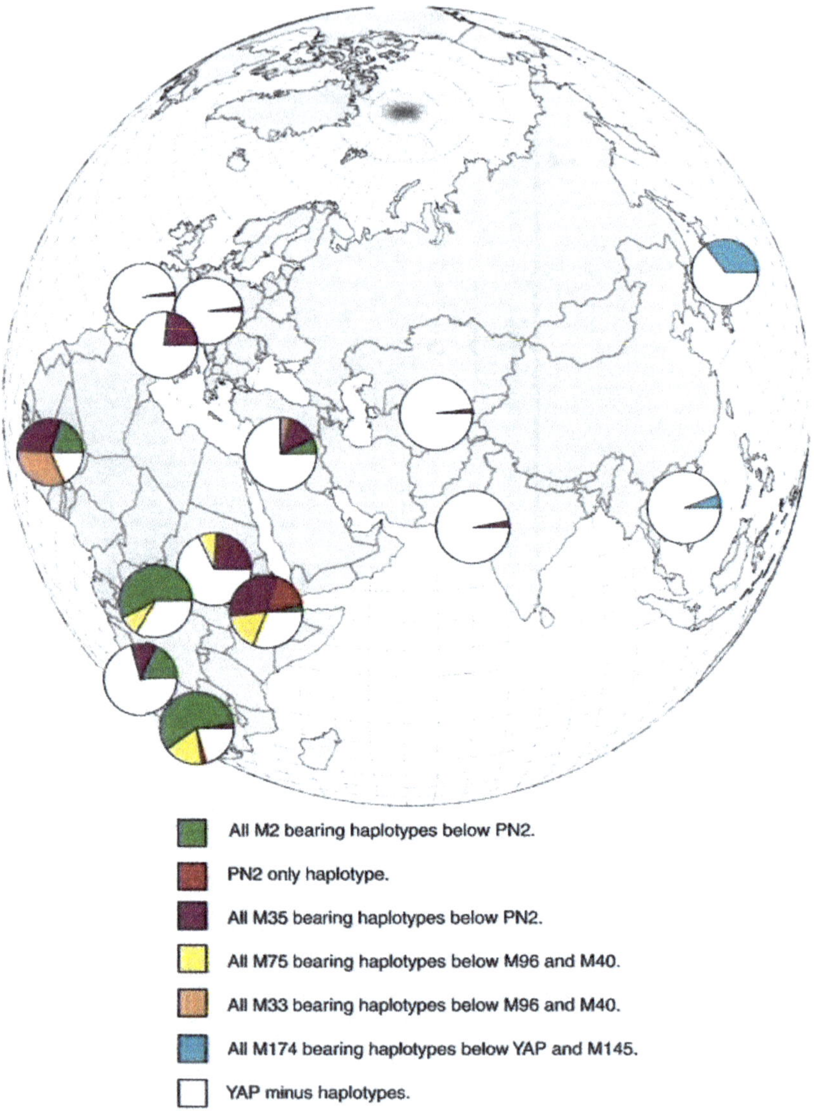

Figure 3. Global frequency distribution of various YAP+/M145A haplotypes in Groups III and IV.

associated with Group III lineages are predominantly African, although European, Central Asian and Pakistani representatives are observed. Conversely, populations associated with Group IV are exclusively localized to the eastern portions of Eurasia.

2. *The out of Africa signature*

The M168 mutation occurs in all lineages contained within Groups III-X. Except for a few occurrences in Europe and West Asia, those lineages lacking the M168 transition represented in Groups I and II are exclusively comprised of individuals from Africa. Groups I and II are on the opposite side of the two most basal splits in the tree from haplogroups III-X. In this respect, the Y phylogeny closely resembles the mtDNA phylogeny (Penny *et al.* 1995). An estimate of the age of the most recent common ancestor based upon coalescence analysis yielded an age of 59,000 years with 95% confidence interval of 40,000 to 140,000 years (Thomson 2000). While the paleoanthropological evidence for the first appearance of modern humans predates this estimate (Lahr and Foley 1994), it falls within the upper limit of the estimate. The M168 mutation is at the root of Groups III-X, which characterize both African and non-African populations. As shown in Figure 2, three clades subsequent to the M168 mutational event persist today which represent the bulk of Y-chromosome diversity outside Africa and provide intriguing clues into the non-African pre-historical pattern of colonizations, differentiations and subsequent migrations overlaid upon previous population ranges. These include the YAP+/M145A pair, RPS4Yand M89.

3. *Synopsis of East Asian Y-chromosome haplogroups*

Groups IV,V,VI, VII and X have been observed in East Asian males (Su et al. 1999, Underhill et al. 2000). Group V lineages are defined by the presence of the RPS4YC711T = M130 mutation (Bergen et al. 1999). Although Group V related lineages are observed in Europe, Siberia, Australasia and the Americas, a subset of haplotypes within Group V also occurs in E. Asians. These include members of populations previously characterized as the null haplotype H1 (Su et al. 1999; Qian et al. 2000). The Asian YAP+ Group IV lineages have been observed in Tibet, Laos and Japan (Underhill et al.1997; Underhill et al. 2000) as well as in Chinese and other East Asian populations characterized as H2 and H3 in Su et al. (1999) and Qian et al. (2000) since they all have M174 (Su, personal communication). The monophyletic Group

VII is defined by M175. Group VII lineages are the most frequently observed in East Asia, some of which are also represented in Polynesia (Su et al. 2000). Group VI lineages are also observed in East Asia at relatively low frequencies which were previously represented as haplotype H4 (Su et al. 1999). Lastly, a single Group X haplotype characterized by mutation M120 has been observed at about 5% frequency in Han populations (Su et al. 1999).

Inferences concerning the relative antiquity of the various Groups in East Asia can be deduced from the Y-phylogeny, haplotype frequency and gene geography. The more recently derived Group VII haplotypes have subsequently overlaid Asian representatives of both Groups IV and V haplotpes which probably represent the descendents of earlier inhabitants. These haplotypes probably have achieved their high frequency due to founder effects and recent population growth catalyzed by rice and millet agriculture (Cavalli-Sforza et al. 1994). Group VI lineages, although at low frequency like the Group IV and V representatives, probably reflect more recent Neolithic migration from West Asia where their frequency is markedly higher, rather than vestiges of early Paleolithic settlers.

4. The YAP+/M145A Group III

The M145 G to A transition is a sequence variant that is phylogenetically paired to the YAP+ mutation. While M145A allele mimics YAP+ it is currently impossible to determine which mutation arose first. The YAP+/M145A lineages within Group III are specifically defined by the presence of the phylogenetically equivalent dual transition mutations M40 (=SRY4064) and M96. Group III lineages are the most frequent in Africa. The PN2 mutation defines most of the haplotypes within Group III. Two major subclades characterize the PN2 lineages that track different microevolutionary events. One is an A to G transition (Seielstad et al 1994) localized within the DYS271 STS and referred to as M2. This mutation is common throughout sub-Saharan Africa with a distribution consistent with the Bantu expansion (Passarino et al. 1998). The other PN2 related subclade is defined by the M35 transversion which occurs in populations from Eastern and Northeast Africa, as well as the Middle East, Europe and Central Asia indicating that M35 lineages participated in demographic events distinctive from the Bantu expansion. Interestedly, some YAP+/M145A lineages do not carry the PN2 mutation (Figure 2) which is suggestive of the persistence of African specific YAP+/M145A lineages that are themselves derived from lineages that are distinctive from the more common and widespread PN2 related lineages.

5. The YAP+/M145A Group IV

The YAP+/M145A lineages within Group IV are all associated with the M174 transition mutation and form a monophyletic clade. The character state marked by the M174 mutation defines the YAP+/M145A lineages that are found in East Asia. As mentioned above, although scattered throughout E. Asia at low frequency with the exception of Tibet and Japan, all YAP+/M145A chromosomes without the M40/M96 mutations have the M174 mutation. The M15 related lineage occurs mostly on the mainland while the M55 related haplotypes are localized to Japan. M174 lineages are observed in E. Asia, which are independent of M15 or M55 related haplotypes. Qian et al. (2000) reported that Tibetan Y chromosomes originated from central Asia (Qian et al. 2000) which is plausible if the region near Mongolia is implied, but not Kazakstan, Kyrgyzstan, etc, since to date only Group III lineages have been found in those locations (Underhill et al. 2000). The presence of Group IV lineages in East Asia (Su et al. 1999; Underhill et al. 2000) and possibly Mongolia as inferred from αh and DYS19 microsatellite data (Santos et al. 1996) represents a geographic dichotomy in which Group IV lineages reflect a more ancient distribution related to Paleolithic Asians, while Group III lineages in Europe, Pakistan and Central Asia may have been caused by Neolithic expansions from the Middle East.

6. Re-evaluating the geographic origin of YAP+

A degree of uncertainty will always exist regarding the geographic origins of DNA sequence variants such as the YAP+ because of a lack of precise knowledge concerning pre-historic spatio-temporal allele distributions. As mentioned above, on the basis of a 9 marker Y-chromosome phylogeny, 1544 samples, nested cladistic and coalesence analyses Hammer et al. (1998) concluded that following an initial expansion of modern humans from Africa to Eurasia, the YAP+ mutational event occurred initially somewhere in Asia about 55,000 ago. This scenario then postulates that YAP+ descendents subsequently participated in a back to Africa migration event. The M40 (=SRY4064) mutation was estimated to have occurred about 31,000 ago and considered to be associated with this bottleneck reflux event. The new Y-chromosome polymorphisms, their phylogenetic context (Underhill et al 2000) and the recent molecular age of Y-chromosome variation provide an opportunity to reappraise the hypothesis of an Asian origin of YAP. Several lines of evidence now tend to contradict the Asian origin scenario. These include:

1. The highest frequency (>80%) of YAP+/M145A lineages occur in Africa and considerable hapltoype differentiation has occurred which is consistent with an African origin. While alternative story lines are possible, they are less probable given the much higher frequency of YAP in Africa relative to Eurasia.

2. The presence of haplotypes at the nodes would be informative because they would provide insights into the ancestral geographic distribution of the YAP+/M145A clade. However the observation that no Y-chromosomes have yet been observed that display one but not the other allele for YAP+ or M145A reduces the interpretative power of phylogeographic approaches. Equally important in this respect are the observations that no Y-chromosomes have yet been detected that are characterized as having the YAP+/M145A pair but without M40, M96 or M174 mutant alleles. This is somewhat unexpected given the recent apparent molecular age of Y chromosome variation and therefore implies a small effective population size and possibly a high turnover rate.

3. While the evidence of a range expansion linking Africa and Asia remains, phylogenetic support for directionality back to Africa based on the presence and absence of character state for M40 is now completely neutralized. Specifically, since M174 acts in an manner analogous to the M40 (=SRY4064)/M96 segment it essentially eliminates any directional support for the back to Africa range expansion based upon the ancestral alleles M40 and M96 being present in Asia and the derived alleles present only in Africa, which was the key factor in the Asian origins hypothesis.

4. While the phylogeography of Group V indicates the defining mutation RPS4YC711T (M130) did in fact arise in Asia at a similar time to that proposed for the YAP+ event, there is no evidence that any Group V lineages migrated back from Africa coincident with the purported YAP+ lineages. Additionally , no other division of the tree shows a back to Africa pattern.

5. There is no paleoanthropological evidence of a back to Africa colonization event (Lahr and Foley 1998).

6. There is no mtDNA or autosomal DNA evidence providing unequivocal directionality regarding a back to Africa migration. To the contrary, mtDNA and autosomes display the opposite pattern (Quintana-Murci et al. 1999, Kidd et al. 1998).
7. While results from a nested cladistic analysis of the new Y data discussed here (Roseman, unpublished results) indicates an out of Africa range expansion over the entire cladogram, there is now no evidence of a "back to Africa" range expansion because any such inferred directionality based upon the previous absence of mutational character state for M40 is now neutralized by the M174 mutation. Furthermore, the nested cladistic analysis of the YAP+ clade did not yield results that could be interpreted as a range expansion.

7. The origin of the YAP+/M145 A branch in the Y chromosome genealogy

Since we cannot determine the individual histories of the YAP+ and M145 G to A mutations, we discuss them jointly. In summary, we propose that the M168 mutation first appeared in Africa 31,000-79,000 years ago (Thomson et al. 2000) on a modern human lineage which preceded extant Groups III-X haplotypes. Subsequently the YAP+/M145A mutations arose in Africa on an undifferentiated M168 haplotype. Some YAP+/M145A descendents remained in Africa and others left to eventually become part of the gene pool representative of the early successful colonizers in Asia. Sometime the M40/M96 mutations appeared in Africa associated with YAP+/M145A lineage and likewise, the M174 mutation occurred independently in Asia on an undifferentiated YAP+/M145A lineage. Finally the precursor YAP+/M145A lineages in both Africa and Asia were completely replaced by the M40/M96 and M174 lineages respectively. Support for the disappearance of earlier haplotypes comes from Group V. The global distribution of RPS4Y lineages implies that an undifferentiated M168 lineage left Africa, then a descendent acquired the RPS4YC711T mutation which subsequently survived in Asia, while undifferentiated M168 lineages vanished or further mutated globally. Until intermediate haplotypes are discovered in either contemporary (assuming they still persist somewhere) or ancient DNA specimens, we contend that this scenario for the African origin of YAP+/M145A is more plausible and parsimonious than the Asian YAP+ origin hypothesis.

Acknowledgements

We thank L. L. Cavalli-Sforza, G. Passarino, M. Jobin and J. Mountain for helpful discussions. This work was supported in part by National Institute of Health Grants GM55279 and HG01707.

References

Altheide, T. K. and Hammer, M. F. 1997. Evidence for a possible Asian origin of YAP+ Y chromomsomes. *Am. J. Hum. Genet.* 61: 462-466.

Bravi, C. M., Baillet, G., Martinez-Marignac, V. L. and Bianchi, N. O. 2000. Origin of YAP+ lineages of the human Y-chromosome. *Am. J. Phys. Anthropol.* 112:149-158.

Bergen, A. W., Wang, C.-Y. , Tsai, J., Jefferson, K., Dey, C., Smith, K. D., Park, S.-C., et al. 1999. An Asian-native American paternal lineage identified by RPS4Y resequencing and by microsatellite haplotyping. *Ann. Hum. Genet* 63:63-80

Cavalli-Sforza, L. L., Menozzi P. and Piazza, A. 1994. *The History and Geography of Human Genes*. Princeton, NJ:University Press, pp. 106-110.

Hammer, M. F. 1994. A recent insertion of an Alu element on the Y chromosome is a useful marker for human population studies. *Mol. Biol. Evol* 11 749-761.

Hammer, M. F. and Zegura, S. L. 1996. The role of the Y chromosome in human evolutionary studies. *Evol. Anthropol.* 5:116-134.

Hammer, M. F., Spurdle, A. B., Karafet, T., Bonner, M.R., Wood, E. T., Novelletto, A., Malaspina, P., et al. 1997. The geographic distribution of human Y chromosomes. *Genetics* 145:787-805.

Hammer, M. F. and Horai, S. 1995. Y chromosomal DNA variation and the peopling of Japan. *Am. J. Hum. Genet.* 56:951-962.

Hammer, M. F., Karafet, T., Rasanayagam, A., Wood, E.T., Altheide, T. K., Jenkins, T., Griffiths, R. C., et al. 1998. Out of Africa and back again: nested cladistic analysis of human Y chromosome variation. *Mol. Biol. Evol* 15:427-441.

Jobling, M. A. and Tyler-Smith, C. 1995. Fathers and sons: The Y chromosome and human evolution. *Trends in Genetics* 11:449-456.

Kidd, K. L., Morar, B., Castiglione, C. M., Zhao, H., Pakstis, A. J., Speed, W., Bonné-Tamir, B., Lu, R-B., et al. 1998. A global survey of haplotype frequencies and linkage disequilibrium at the DRD2 locus. *Hum. Genet.* 103:211-227.

Lahr, M. M. and Foley, R. 1994. Multiple dispersals and modern human origins. *Evol. Anthropol.* 3:48-60.

Lahr, M. M. and Foley, R. 1998. Towards a theory of modern human origins: geography, demography and diversity in recent human evolution. *Yrbk of Phys. Anthropol.* 41:137-176.

Oefner, P. J. and Underhill, P. A. 1998. *Current Protocols in Human Genetics*. New York: Wiley and Sons, Supplement 19,7.10.1-7.10.12.

Passarino, G., Semino, O., Quintana-Murci, L., Excoffier, L., Hammer, M. and Santachiara-Benerecetti, A.S. 1998. Different genetic components in the Ethiopian population, identified by mtDNA and Y-chromosome polymorphisms. *Am J Hum Genet*. 62:420-434.

Penny, D., Steel, M., Waddell, P. J. and Hendy, M. D. 1995. Improved analyses of human mtDNA sequences support a recent African origin for Homo sapiens. *Mol. Biol. Evol*. 12:863-882.

Qian, Y., Qian, B., Su, B., Yu, S., Ke, Y., Chu, Z., Shi, L., Lu, D., Chu, J. and Jin, L. 2000. Multiple origins of Tibetan Y chromosomes. *Hum. Genet*. 106: 453-454

Quintana-Murci, L., Semino, O., Bandelt, H-J., Passarino, G., McElreavey, K. and Santachiara-Benerecetti, A. S. 1999. Genetic evidence of an early exit of Homo Sapiens sapiens from Africa through eastern Africa. *Nat. Genet*. 23:437-441

Santos, F. R., Bianchi, N. O. and Pena, S. D. J. 1996. Worldwide distribution of human Y-chromosome haplotypes. *Genome Res*. 6:601-611.

Seielstad, M. T., Hebert, J. M., Lin, A. A., Underhill, P. A., Ibrahim, M. D., Vollrath, D., and Cavalli-Sforza L. L. 1994. Construction of human Y-chromosomal haplotypes using a new polymorphic A to G transition. *Hum. Mol. Genet*. 3:2159-2161.

Shen, P., Wang, F., Underhill, P.A., Franco, C., Yang, W-H, Roxas, A., Sun, R., et al. 2000. Population genetic implications from sequence variation in four Y chromosome genes. *Proc. Natl. Acad. Sci. USA* 97:7354-7359

Spurdle, A. M., Hammer, M. F. and Jenkins, T. 1994. The Y Alu polymorphism in southern African populations and its relationship to other Y-specific polymorphisms. *Am J Hum Genet* 54:319-330.

Su, B. 2000. Personal communication. Email dated 30 Jan 2000.

Su, B., Xiao, J., Underhill, P., Deka, R., Zhang, W., Akey, J., Huang, W., et al. 1999. Y-chromosome evidence for a northward migration of modern humans into eastern Asia during the last ice age. *Am. J. Hum Genet*. 65:1718-1724.

Su, B., Jin, L., Underhill, P. A., Martinson, J., Saha, N., McGarvey, S. T., Shriver, M. D., et al. 2000. Polynesian origins: insights from the Y chromosome. *Proc. Natl. Acad. Sci. USA* 97:8225-8228.

Thomson, R., Pritchard, J. K., Shen, P., Oefner, P. J. and Feldman, M. W. 2000. Recent common ancestry of human Y chromosomes: evidence from DNA sequence data. *Proc. Natl. Acad. Sci. USA*. 97:7360-7365.

Underhill, P. A., Shen, P., Lin, A. A., Jin, L., Passarino, P., Yang, W. H., Kauffman, K., et al. 2000. Y chromosome sequence variation and the history of human populations. *Nature Genet*. 26:358-361.

Underhill, P.A., Jin, L., Lin, A. A., Mehdi, S. Q., Jenkins, T., Vollrath D., Davis, R.W., et al. 1997. Detection of numerous Y chromsome biallelic polymorphisms by denaturing high performance liquid chromatography. *Genome Res*. 7:996-1005.

Underhill, P. A., Passarino, G., Lin, A. A., Shen, P., Foley, R. A., Mirazón Lahr, M., Oefner, P. J. and Cavalli-Sforza, L. L. 2001. The phylogeography of Y chromosome binary haplotypes and the origins of modern human populations. Ann. Hum. Genet. 65: 43-62.

Whitfield, L. S., Sulston, J. E. and Goodfellow, P. N. 1995. Sequence variation on the human Y chromosome. *Nature* 378:379-380.

GENETIC HISTORY OF ETHNIC POPULATIONS IN SOUTHWESTERN CHINA

BING SU
Kunming Institute of Zoology, the Chinese Academy of Sciences
Human Genetics Center, University of Texas

CHUNJIE XIAO
Department of Biology, Yunnan University

LI JIN
Human Genetics Center, University of Texas

Introduction

Yunnan province is located at the Himalaya/Heng-Duan mountain region of Southwestern China and encompasses an extremely diversified ecological environment, from tropical rain forest in the south to Frigid Zone conifers in the northwest. Besides the rich biodiversity of plant and animal species, it is also the place where many ethnic cultures have coexisted during the past several thousand years. Therefore, elucidation of the prehistoric migrations that led to the peopling of Yunnan becomes fascinating. There are 26 officially recognized ethnic populations currently living in Yunnan. They are Han Chinese, Yi, Bai, Hani, Dai, Zhuang, Miao, Lisu, Hui, Lahu, Wa, Naxi, Yao, Jingpo, Tibetan, Blang, Buyi, Pumi, Achang, Nu, Jinuo, Deang, Mongolian, Shui, Manchurian and Dulong (Figure 1).

According to historical records, people came to Yunnan at different times from different parts of China. About 2,000 years ago, there were three major ancient population lineages: Bai-Yue, Bai-Pu and Di-Qiang (Wang 1994). The Bai-Yue and Bai-Pu were the two major ancient population groups living in southern China and related to the He-Mu-Du culture, which occurred about 7,000 years ago in the Mid-Lower Yangtze River basin. The word *Bai* in Chinese means *hundreds*. Therefore, Bai-Yue and Bai-Pu are in fact the common names of multiple sublineages within the same group. According to the Neolithic archaeological findings, Bai-Yue and Bai-Pu have been living in Yunnan for the past 4,000 years and might be the earliest modern human inhabitation in that area (Cang 1997). The Bai-Yue and Bai-Pu were the dominant populations in the southern and eastern parts of China

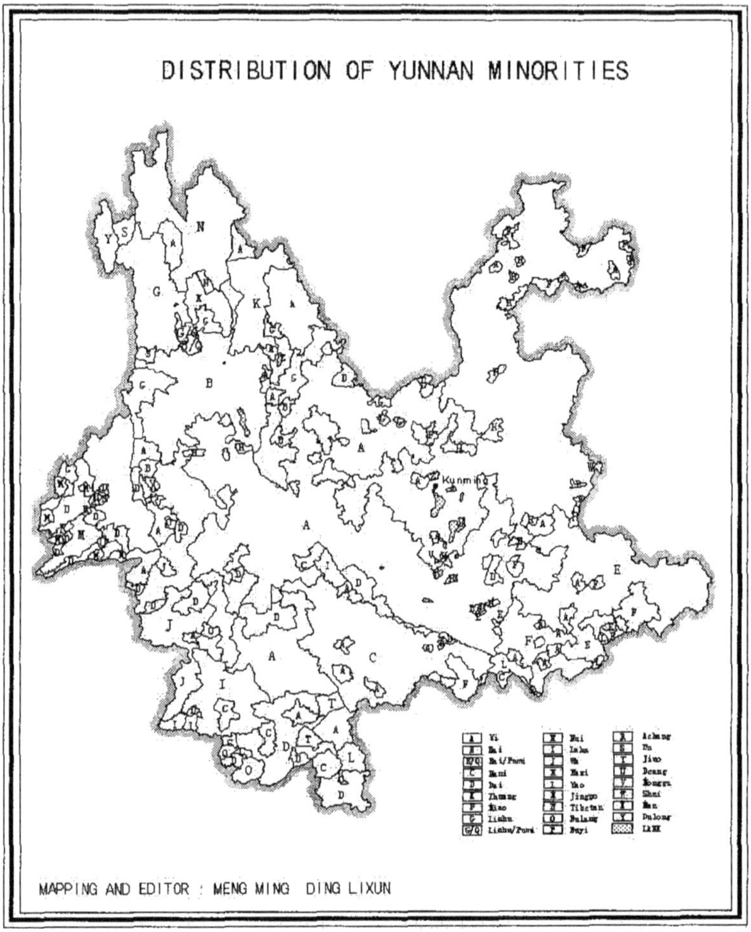

Figure 1. Distribution of Yunnan minorities.

(Cang 1997). Among the current Yunnan ethnic populations, Dai, Shui and Zhuang are the descendants of Bai-Yue while Wa, Deang and Blang are the descendants of Bai-Pu. Di-Qiang was a northern population group related with the Yang-Shao culture which occurred about 8,500 years in the Upper-Mid Yellow River basin (Wang 1994; Cang 1997). Literary records show that the major migrations of Di-Qiang to Yunnan happened during the Qin Dynasty about 2,700 years ago (Cang 1997). They came from the north, by way of the well-known *Zang (Tibet)-Mien Corridor* starting at the Upper Yellow River region. The current Yunnan descendants of the Di-Qiang lineage include Yi, Bai, Hani, Lisu, Lahu, Naxi, Jingpo, Tibetan, Buyi, Pumi, Achang, Nu, Jinuo and Dulong.

Linguistically, Bai-Yue, Bai-Pu and Di-Qiang correspond with the Daic, Austro-Asiatic and Tibeto-Burman subfamilies of the Sino-Tibetan language family, respectively (Wang 1994; Wang 1995; Grimes 2000). Besides being descendant populations of the three ancient lineages, the other ethnic populations including Han Chinese came to Yunnan relatively recently (Cang 1997). The earliest records of Miao/Yao in Yunnan can be dated back to the Han Dynasty about 1,000 years ago. However, the major migration of Miao/Yao to Yunnan happened only about 200 years ago in the Qing Dynasty. Linguistically, the Miao/Yao belong to the Hmong-Mien family. The Mongolian and Hui came to Yunnan about 750 years ago in the Song Dynasty while the Manchurian migration occurred in the Qing Dynasty about 200 years ago. These three populations speak languages of the Altaic family, the dominant language family in northern Asia. The major Han Chinese migrations to Yunnan started recently in the Ming Dynasty about 600 years ago according to the literature. Han Chinese belongs to the Chinese subfamily of the Sino-Tibetan language family (Cang 1994).

With multiple populations living in the same region, population admixture usually leads to extensive genetic exchange among different ethnic populations. However, due to the complex landforms with high mountains and big rivers, instead of a melting pot, Yunnan serves as a mosaic of land pieces for different ethnic groups with a certain degree of geographic isolation. Consequently, many Yunnan ethnic populations have kept their cultural traditions and social structures along with their history. For example, the Muo-Suo people have been keeping their ancient matriarchal social system during the past several thousand years. Therefore, it would be interesting to compare the genetic background of the different ethnic populations with the culture diversification related to the historical migrations in Yunnan. In this preliminary study, using both Y chromosome and mitochondrial DNA markers, 13 Yunnan ethnic populations were analyzed in order to retrieve the origins of those ethnic populations and the extent of population admixture among them.

Materials and Methods

A total of 536 DNA samples were collected, covering the descendant populations of the three main ancient lineages (Table 1). Among them, 275 samples are males and were analyzed for the Y chromosome markers. There are 13 populations representing three different language families, including Dai (Daic family), Wa, Deang, Blang (Austro-Asiatic family), Tibetan, Jingpo, Akha, Lahu, Yi, Jinuo, Bai, Naxi and Han Chinese (Sino-Tibetan family).

The 19 Y chromosome biallelic markers were chosen based on their population specificity, including M1 (Alu insertion, also called YAP), M3 (C to T substitution), M5 (A to G substitution), M7(C to G substitution), M9(C to G substitution), M15 (9bp insertion), M17 (1bp deletion), and M45 (G to A substitution), M50 (T to C substitution), M88 (A to G substitution), M89 (C to T substitution), M95 (C to T substitution), M103 (C to T substitution), M110 (T to C substitution), M111 (4bp deletion), M119 (A to C substitution), M120 (T to C substitution), M122 (T to C substitution), and M134 (1bp deletion) (Vollrath et al. 1992; Hammer et al. 1997; Underhill et al. 1996, 1997; Su et al. 1999). M1 is an ancient polymorphism which is found in both African and Asian populations, but is generally absent in other populations (Hammer et al. 1997). M3 is an American Indian specific marker while M5 is Oceanian specific (Underhill et al. 1996, 1997). M45 has a high frequency in Caucasians, but is relatively rare in East Asian populations (Su et al. 1999). M9 is generally absent in Africans, but frequent in non-African populations (Underhill et al. 1996; Su et al. 1999; Underhill et al. 2000). M122, M95 and M119 define three lineages which are found predominantly in East Asian populations but are absent in other parts of the world (Su et al. 1999). An allele specific genotyping assay was used to type the 19 Y chromosome biallelic markers following our previous report (Su et al. 1999).The mtDNA B-locus (9bp deletion) was also typed in the 536 samples using previously reported protocols (Passarino et al. 1993).

Table 1. Frequency distribution of Y chromosome haplotypes and mtDNA B-locus polymorphism in 13 Yunnan populations

Population	Language	Size	H1	H2	H3	H4	H5	H6	H7	H8	H9	H11	H12	H13	H14	B locus	Size
	Most recent marker associated		M130	M1	M15	M89	M9	M122	M7	M134	M119	M95	M88	M120	M45		
Dai	Daic	40	5			7.5	23	13		30		5	18			21.99	141
Wa	Austro-Asiatic	47	8.5			2.1	21	32		34		2.1				6	50
Deang	Austro-Asiatic	2							50	50						40	10
Blang	Austro-Asiatic	24	17				17	17		13		38				16	50
Tibetan	Sino-Tibetan, Tibeto-Burman, Bodic	27		14.8	29.6	3.7	14.8	7.4		29.6						10	20
Jingpo	Sino-Tibetan, Tibeto-Burman, Baric	5								100						0	17
Akha (Aini)	Sino-Tibetan, Tibeto-Burman, Burmese-Lolo	3						33						33	33	2.04	49
Lahu	Sino-Tibetan, Tibeto-Burman, Burmese-Lolo	37	8.14			18.9	37.8	16.2		10.8		8.09				2.86	35
Yi	Sino-Tibetan, Tibeto-Burman, Burmese-Lolo	14			14		43	21		7.1		14				21.21	20
Jino	Sino-Tibetan, Tibeto-Burman, Burmese-Lolo	18	17			5.6	39	11		22		5.6				5.3	19
Bai	Sino-Tibetan, Tibeto-Burman, Burmese-Lolo	13	15		7.7		15	31		31						13.33	30
Naxi	Sino-Tibetan, Tibeto-Burman, Burmese-Lolo	18					78					17			5.5	25.53	47
Han	Sino-Tibetan, Chinese	27	11.1			3.7	3.7	18.5		55.6	3.7	3.7				18.75	48

Results and Discussion

The haplotypes of 19 Y chromosome biallelic markers in a total of 275 male individuals from 13 Yunnan populations are shown in Table 1. A total of 17 haplotypes were observed in our previous study on the extent East Asian populations and no recurrent mutations were identified (Su et al. 1999). The phylogenetic relationship among the haplotypes is shown in Figure 2 of Chapter 9. Of the 17 haplotypes, 13 haplotypes are present in the Yunnan populations. The frequency distribution of the 13 haplotypes is listed in Table 1. Haplotypes H5, H6, H8 and H11 are the predominant ones in most of the populations showing their East Asian affinity since H6-H12 are East Asian specific haplotypes and are not present in other world populations. H1 and H4 are two relatively ancient haplotypes defined by M130T and M89C, respectively. They have medium frequencies in some of the populations.

Our previous study of the 19 Y chromosome markers on extant East Asian populations showed that East Asians have a common origin in Africa and about 60,000 years ago, they migrated to the southern part of East Asia and then expanded northward to the other parts of East Asia (Su et al. 1999, 2000a; Jin and Su 2000). However, due to migration and genetic drift, populations in the same language families tend to show similar signatures of haplotype distributions, which are distinctive from other language families. There are six language families in East Asia, besides the aforementioned three families in this study, Hmong-Mien, Austronesian and Altaic are the other three families (Grimes 2000). The Altaic speaking populations are mainly located in northern East Asia, mostly in Siberia. The other two families are spoken in southern China, Southeast Asia and the Pacific Islands.

In Sino-Tibetan populations, sharing a T to C mutation at locus M122, H6, H7 and H8 are the predominant haplotypes in almost all of the Sino-Tibetan populations (Su et al. 2000b). The M122C allele is East Asian specific and absent in Africans, Europeans and Oceanians as revealed in our previous study (Su et al. 1999). Recent studies in the extant Siberian and Central Asia populations only revealed the sporadic occurrence of M122C and M119C (H9) (Su et al. 2000b). H6 is the ancestral haplotype of the M122C alleles, while H7 and H8 are the two derived ones with additional mutations, M7 and M134, respectively. The average frequency of M122C in Sino-Tibetan populations is 52.2% (28.6-100%) the highest among the six language families in East Asia (Hmong-Mien 47.6%, Daic 26.1%, Austronesian 25.7%, Altaic 24.5%, and Austroasiatic 21.8%) (Su et al. 2000b). Therefore, the prevalence of M122C, especially the high frequency of H6 and H8, can be taken as a genetic signature for the Sino-Tibetan populations (Su et al. 2000b). Another feature of Sino-Tibetan populations is the total absence of H10 and the absence of H9 in Tibeto-Burman subfamilies. H2 and H3 (defined by YAP+) are the characteristic haplotypes for the Tibetans and they originated in Central Asia/Southwest Siberia (Qian et al. 2000).

In Daic and Austro-Asiatic populations, the Y chromosome profiles are extremely diversified, covering all the East Asian specific haplotypes (H6-H12) (Su et al. 1999). The Daic and Austro-Asiatic populations might be the most ancient people following the initial settlement of modern humans of African origin tens of thousand years ago in southern East Asia. H11 and H12 (defined by M95T) are the most dominant haplotypes in these populations. M122C haplotypes are also present in most of the populations with medium frequencies. The antiquity of Daic and Austro-Asiatic populations is also reflected by the within family divergence of haplotype distributions in different populations.

In the 13 Yunnan populations studied, the nine Sino-Tibetan populations, in general, fit into the patterns described above showing a high frequency of M122C haplotypes, except the Naxi. Hence, this genetic pattern confirms the literary records about the northern origin of the Sino-Tibetan populations in Yunnan, who are descendants of the ancient Di-Qiang people. Interestingly, populations in different language branches of the Sino-Tibetan family tend to show some distinctive haplotype distributions (Su et al. 2000b; Martisoff 1991). The Yunnan Tibetan (Bodic branch) has a very similar haplotype distribution with the Tibetans in Tibet, with the significant occurrence of YAP+ haplotypes (H2 and H3). The Jingpo people (Baric) are fixed for H8, one of the derived M122C haplotypes though the sample size is small. Our study on multiple Baric populations from northeastern India indicated that the dominant occurrence of H8 is a common feature in this language branch (Su et al. 2000b). The Baric populations have extremely homogeneous Y chromosomes. The M122C/M134G (H8) alleles are almost fixed in this language branch with, only the sporadic occurrence of other Y haplotypes. The extremely high frequencies of H8 (84.4% on average) in all the Baric populations strongly suggests a population bottleneck event probably associated with the branching of Baric from the ancestral Proto-Tibeto-Burman population (Su et al. 2000b).

The occurrence of H11 in the 6 Yunnan Burmese-Lolo (a language branch of Tibeto-Burman) populations reflects an extensive population admixture between Burmese-Lolo and Daic and Austro-Asiatic populations. This is consistent with the current distribution (Figure 1) and historic migrations of Burmese-Lolo populations both in Yunnan and then from Yunnan to Southeast Asian countries, such as Myanmar, Vietnam and Thailand (Matisoff 1991; Su et al. 2000b). Compared with Tibetan and Jingpo who are only found in northwestern Yunnan, the Burmese-Lolo populations have a much wider distribution from north to south, neighboring the Daic and Austro-Asiatic populations. However, admixture was not observed in the other two Tibeto-Burman branches with H11 totally absent in the Bodic (Yunnan-Tibetan) and Baric (Jingpo) populations. Surprisingly, Akha seems to be different from other Burmese-Lolo populations by having the M45A haplotypes (H13 and H14). This unique feature was confirmed in another Akha population (13 individuals tested) from Thailand (H13, 69.2%). Since M45A haplotypes are prevalent in European, Central Asia and Indian populations, the Akha populations could have had a strong influence from Caucasian populations in history. Furthermore, under the background of M45A, H13 has another polymorphism defined by M120C that is only observed in the Akha and Han Chinese (2.2%), but is absent in other East Asian and world populations (Su et al. 1999, 2000a, 2000b; Qian et al. 2000; Underhill et al. 2000). Therefore, the M120C mutation might occur in East Asia.

The H9 are absent in all the Sino-Tibetan populations except in the Yunnan Han Chinese. Hence, this eastern-distributed haplotype (which shows up in Altaic, Han Chinese, eastern Daic and Austronesian populations) does not have much influence on the Yunnan ethnic populations. Our previous study showed that in the Hmong-Mien populations, H7 is one of the dominant haplotypes defined by M7G (28.6% on average) (Su et al. 2000b). Except one individual in the Deang population, H7 is absent in all the Yunnan populations studied, which implies a minor contribution of Hmong-Mien to Sino-Tibetan, Daic and Austro-Asiatic, due to their short history and isolated distribution in Yunnan (Cang 1997).

The mtDNA 9-bp deletion (B-locus) distribution of 536 individuals is shown in Table 1. It is not quite structured in the Yunnan populations studied. However, the B-locus data in extant East Asian populations showed that southern populations tend to have a higher frequency of the 9-bp deletion than that of northern populations (unpublished data; Yao et al. 2000). This contrast is not well reflected in the Yunnan populations. The presumed northern originated Tibeto-Burman populations, Jingpo, Akha Jinuo and Lahu have low frequencies while Tibetan, Yi, Bai and Naxi have frequencies ranging from 10%-25%. On the other hand, the four presumed southern populations, Dai, Deang and Blang seem to fit the north/south contrast with high frequencies of the B-locus polymorphism, but leaving the Wa as an exception (6%). Interestingly, the low frequency of the B-locus polymorphism in the Wa is quite consistent with its Y chromosome haplotype distribution. The M122C haplotype frequency in the Wa is as high as 65.9%, much higher than that in other southern populations. Considering its well-diversified haplotypes (6 in total), the high frequency of M122C in Wa is more likely a reflection of the strong influence of Sino-Tibetan populations, instead of the consequence of within-population genetic drift. In general, the unstructured distribution of the B-locus polymorphism might implicate higher female gene flow, usually observed in geographically closely related populations (Seielstad et al. 1998).

In summary, our data on Y chromosome and mtDNA polymorphisms of Yunnan ethnic populations are consistent with literary records about the three ancient population lineages, Bai-Yue, Bai-Pu and Di-Qiang. As the descendants of Di-Qiang, the Sino-Tibetan populations in Yunnan still have the genetic signature of high frequency of M122C haplotypes, confirming the northern origin of those populations in Yunnan. The Daic and Austro-Asiatic populations, the descendants of Bai-Yue and Bai-Pu are ancient populations with diversified Y chromosome haplotypes and consistent with an early southern origin in the literary record. In addition, as indicated by both Y chromosome and mtDNA polymorphisms, extensive population admixture occurred between southern and northern populations in Yunnan, though geographic barriers exist.

Acknowledgement

B.S. and L.J. were supported by NIH grants. CJ.X. was supported by a grant from National Natural Science Foundation of China.

References

Cang, M. 1997. *Culture and Migration History of Yunnan ethnic populations.* Kunming:Yunnan Ethnic Press (in Chinese).

Grimes, B.F. (ed). 2000. *The Ethnologue: Languages of the World.* Summer Institute of Linguistics: http://www.sil.org/ethnologue/.

Hammer, M.F., Spurdle, A.B., Karafet, T., Bonner, M.R., Wood, E.T., Novelletto, A., Malaspina, P., et al. 1997. The geographic distribution of human Y chromosome variation. *Genetics* 145:787-805.

Jin, L. and Su, B. 2000. Natives or immigrants: Modern human origin in East Asia. *Nature Reviews Genetics* 1:126-133.

Martisoff, J.A. 1991. Sino-Tibetan linguistics: Present state and future prospects. *Ann. Rev. Anthropol.* 20: 469-504.

Passarino, G., Semino, O., Modiano, G., Santachiara-Benerecetti, A.S. 1993. COII/tRNA^{Lys} intergenic 9-bp deletion and other mtDNA markers clearly reveal that the Tharus (southern Nepal) have oriental affinities. *Am. J. Hum. Genet.* 53:609-618.

Qian, Y.P., Qian, B.Z., Su, B., Yu, J.K., Ke, Y.H., Chu, Z.T., Shi. L., et al. 2000. Multiple origins of Tibetan Y chromosomes. Human *Genetics* 106: 453-454.

Seielstad, M. Minch E., Cavalli-Sforza L. L. 1998. Genetic evidence for a higher female migration rate in humans. Nat. Genet. 20:278-280.

Su,B., Xiao, J.H., Underhill, P., Deka, R., Zhang, W.L., Akey, J., Huang, W., et al. 1999. Y Chromosome evidence for a northward migration of modern humans in East Asia during the Last Ice Age. *Am. J. Hum. Genet.* 65:1718-1724.

Su B., Jin L., Underhill, P., Martinson, J., Saha, N., McGarvey, S.T., Shriver, M.D., et al. 2000a. Polynesian origins: Insights from the Y chromosome. *Proc. Natl. Acad. Sci. USA* 97: 8225-8228.

Su, B., Xiao, C.J., Deka, T., Seielstad, M., Kangwanpong, D., Xiao, J., Lu, D.R. et al. 2000b. Y chromosome haplotypes reveal prehistorical migrations to the Himalayas. *Hum. Genet.* 107:582-590.

Underhill, P.A., Jin, L., Zemans, R., Oefner, P.A., Cavalli-Sforza, L.L. 1996. A pre-Columbian human Y chromosome-specific transition and its implications for human evolution. *Proc. Natl. Acad. Sci. USA* 93:196-200.

Underhill, P.A., Jin, L., Lin, A.A., Mehdi, S.Q., Jenkins, T., Vollrath, D., Davis, R.W., et al. 1997. Detection of numerous Y chromosome biallelic polymorphisms by denaturing high-performance liquid chromatography. *Genome Res.* 7:996-1005

Underhill, P.A., Shen, P., Lin, A.A., Jin, L., Passarino, G., Yang, W.H., Kauffman, E., et al. 2000. Y chromosome sequence variation and the history of human populations. *Nat. Genet.* 26:358-361.

Vollrath, D., Foote, S., Hilton, A., Brown, L.G., Beer-Romero, P., Bogan, J.S., Page, D.C. 1992. The human Y chromosome: A 43-interval map based on naturally occurring deletions. *Science* 258:52-59.

Wang, Z.H. 1994. *History of Nationalities in China.* Beijing: China Social Science Press.

Wang, W.S.Y. 1995. The ancestry of Chinese: Retrospect and Prospect. *J. Chinese Linguistics Monograph* No. 8.

Yao, Y.G., Watkins, W.S., Zhang, Y.P. 2000. Evolutionary history of the mtDNA 9-bp deletion in Chinese populations and its relevance to the peopling of east and southeast Asia. *Hum. Genet.* 107:504-512.

Y-CHROMOSOMAL VARIATION IN UXORILOCAL AND PATRILOCAL POPULATIONS IN THAILAND

METAWEE SRIKUMMOOL
Department of Biology, Chiang Mai University

DAOROONG KANGWANPONG
Department of Biology, Chiang Mai University

NADIA SINGH
Program for Population Genetics, Harvard School of Public Health

MARK SEIELSTAD
Program for Population Genetics, Harvard School of Public Health

Abstract

The Y chromosome, like mitochondrial DNA, is inherited in a sex-specific manner. This allows a variety of insights into sex-specific differences among populations. One such difference, which has attracted some attention, is the potential asymmetry in male versus female migration rates. Cultural variation in the rules of postmarital residence is expected to influence migration rate differences between the sexes. Here, we report on Y-chromosomal data from four populations inhabiting northern Thailand that follow differing rules of postmarital residence. Variation on the paternally inherited Y chromosome is highest in the Karen, an uxorilocal population — consistent with a higher male than female migration rate as expected under uxorilocality. Data from the maternally transmitted mitochondrial genome will be added in future studies, to more precisely quantify rates of female versus male migration. A survey of variation in the X-chromosomal G6PD gene revealed two previously characterized variants in the Karen and Yao (Iu Mien) populations.

Introduction

From the dawn of the DNA era, human population geneticists have been fascinated by the mitochondrial genome. Heretofore, its most attractive feature has been an apparent inability to undergo genetic recombination. It shares this property with the male-specific portion of the Y chromosome, which has only more recently come under study. The lack of recombination means that all modern mtDNA and Y-chromosomal sequences descend from a single ancestral molecule at some point in the past (a feature known as 'coalescence'), and that a modern non-recombining DNA sequence will differ from an ancestral molecule only by the accumulation of mutations — a process which appears to be fairly regular when viewed over appreciable evolutionary periods. Thus, it has been a relatively straightforward task to assess the extent of genetic variation in living humans and to estimate the time required to generate it. Early students of mitochondrial variation recognized that our origins as a species should not antedate the coalescence time of mtDNA (barring a very complicated and unlikely history of natural selection and migration) (Cann et al. 1987; Vigilant et al. 1991). Estimates of the Y chromosome's coalescence have only just been published, and the dates — though more recent — are in general agreement with the results from mtDNA (Underhill et al. 2000). From these coalescent studies, we have learned that our species is not particularly old in evolutionary terms (almost certainly less than 150,000 years, and perhaps substantially 'younger') and that we, like our closest primate relatives, originated in Africa.

But it is not only their lack of recombination that distinguishes mtDNA and the Y chromosome from the autosomes. A possibly more arresting feature is their sex-specific mode of inheritance. Through its involvement in sex determination, the Y chromosome is paternally transmitted by necessity. In contrast, mtDNA is maternally transmitted, because the mitochondrion resides in the cytoplasm, whose vast bulk is provided by the mother's egg cell.

The sex-specific genetics of mtDNA and the Y chromosome are just beginning to be exploited (Poloni et al. 1997, Seielstad et al. 1998, Perez-Lezaun et al. 1999, Carvajal-Carmona et al. 2000, Jorde et al. 2000, Mesa et al. 2000, and Seielstad 2000). Early analyses indicated that populations exhibit a high degree of population substructure (a high Fst) for the Y chromosome (Poloni et al. 1997 and Seielstad et al. 1998). When compared to a relatively lower degree of population substructure for mtDNA, it was suggested that gene flow (i.e., migration) among most human populations has been mediated primarily by women (Seielstad et al. 1998). Populations are generally more similar to one another from the viewpoint of mtDNA than they are from a Y-chromosomal perspective. Thus women appear to have a higher migration rate than men in most societies, shuttling mtDNA among (sub)populations more frequently than the Y-chromosomal cargo of men. However surprising this result may initially seem, it is easily understood in terms of the rules of postmarital residence that operate in the majority of traditional societies. Most human populations are patrilocal, meaning that, soon after marriage, a wife will typically move from her family into the natal household of her husband. Men, whatever their movements over a lifetime, will generally return to and rear their children at the place of their birth. This is especially pronounced when land and other forms of non-portable wealth are paternally inherited.

Although a higher female migration rate may be the most likely explanation for the generally greater degree of population substructure exhibited by the Y chromosome, additional factors may contribute to or explain the entire discrepancy. High levels of polygyny (as the most frequent cause of greater variance in male vs. female reproductive output — i.e., a smaller male vs. female effective population size) could have a similar effect, since drift (and thus the tendency for populations to differentiate) would be accentuated by the smaller Y-chromosomal N_e. But this greater variance in male reproductive output may not have such a large effect, because it will also reduce the autosomal N_e (Nunney 1991, 2000) and because polygyny — as it is actually practiced throughout the world — often involves maternal relatives of the first wife who would share mitochondrial sequences and up to half of their autosomes.

Unfortunately, standard models of population structure cannot by themselves distinguish between differences in male vs. female effective population size (N_e) and migration rate differences. The classic 'island model' of migration can only relate F_{ST} (the parameter we are able to estimate from genetic data) to the *product* of N_e and migration, with no weighting given to their individual contributions [$F_{ST}=1/(1+4N_em)$, where m is the migration rate among populations]. Because current analytical tools are unable to distinguish between differences in migration rates or effective population sizes, the relative contributions of these two factors to modern patterns of population substructure might best be estimated by comparing several populations that differ in their extent of polygyny and their rules of postmarital residence.

Not all populations are patrilocal, and some follow other patterns of postmarital residence. The Karen of Northwestern Thailand (and Northeastern Myanmar [Burma]) are predominantly uxorilocal (a husband generally resides with the family of his wife), and the patterns of Y-chromosomal and mitochondrial variation are expected to differ in these populations relative to their patrilocal neighbors. The goal of the ongoing study reported here is to assess the patterns of Y vs. mitochondrial variation in several populations of northern Thailand, with a particular eye toward the ways in which cultural variation in the rules of postmarital residence affects the geographic distribution of Y-chromosomal and mitochondrial variation. The available Y-chromosomal data — in showing greater diversity in the Karen relative to surrounding populations — are consistent with a higher male migration rate in the Karen, but there are additional explanations for the result, which cannot be eliminated without (forthcoming) data from additional loci — notably mtDNA.

The haplodiploid inheritance of the X chromosome (males have one copy while females have two), implies that the same sexual asymmetries we have just been discussing will also affect variation on this chromosome. A male migrant will only bring one X chromosome to a population, and will transmit it to only ½ of his offspring; a female migrant, on the other hand, will bring two X chromosomes to a population, and will transmit one of them to each of her children. Given the considerable size and gene density of the X, it is conceivable that (via effects on X-chromosomal N_e), sexual asymmetries in migration rates and N_e will also alter the selective environment of genes that reside on it (the same could be said of mtDNA and the Y chromosome, but their relative gene paucity may limit opportunities for natural selection). In small populations, only alleles with the most dramatic effects on fitness will overcome the effects of drift. In general, only alleles for which $N_e s > 1$ (where s is the relative fitness conferred by the allele) will be driven predominantly by their selective effects, and not primarily by population size (i.e., by genetic drift) (Kimura 1962). If these various cultural practices did significantly alter the X-chromosomal N_e and, thus, the selective environment of X-chromosomal alleles, then the cultural practices themselves would have quantifiable fitness effects, via their effects on biological selection at the DNA level. We have sought evidence in support of this hypothesis by studying the evolution of G6PD, an X-chromosomal gene, whose deficiency alleles are known to confer some resistance to infection with malarial parasites (Ruwende et al. 1995). The data currently available are far too sparse to illuminate this highly speculative hypothesis — even in the limited case of G6PD's evolution. Indeed, the effect of cultural variation on the X chromosome's N_e may be too slight to detect such an effect — should one exist — with realistic sample sizes. What we do show in a very small sample is that the uxorilocal Karen have the highest frequency of G6PD deficiency alleles.

Materials and Methods

Populations

Karen: 21 Skaw Karen males from the Mae Sawan Noi village, Mae Sariang district, Mae Hong Son province. The village was established 130 years ago.

Hmong: 29 males from the Chiang Mai province. 12 are Black Hmong from the Mae Khi village, Mae Rim district and from Doi Pui village, Muang district (2 samples). The other 17 are White Hmong from the Pa Nok Kok village (5 samples) and the Pang Loong village (12 samples). Both villages are in the Mae Rim district. These two subgroups moved from neighboring areas and have been settled in their current villages for 40-45 years.

Akha: 15 males from the Ahjah village, Muang district, Chiang Rai province. This community moved from the Mae Jun district of the same province around 45 years ago.

Lisu: 12 males from the Wiang Klang village, Muang district, Chiang Rai province. They moved from the Mae Jun and Mae Sruay districts, Chiang Rai province 21 years ago.

Laboratory Methods

Total genomic DNA was extracted from whole blood samples from 4 ethnic groups living in northern Thailand — 21 Karen, 29 Hmong, 15 Akha and 12 Lisu — according to a standard 'salting out' protocol described in (Seielstad et al. 1999). Three Y-specific tetranucleotide repeat loci (DYS19, DYS393 and DYS389) were amplified. For each 50 μl PCR volume, 100 ng of total DNA in 800 μM of dNTPs, 3.5 mM $MgCl_2$, 0.2 μM of the two oligonucleotide primers, 5 μl of 10X PCR buffer and 0.7 units of AmpliTaq Gold^{TM} Polymerase (Perkin-Elmer), were thermal cycled. PCR cycling condition for these loci were: an initial denaturation step of 95^{0}C for 10 minutes; followed by 30 cycles of 94^{0}C for 1 minute, 54^{0}C for 1 minute, 72^{0}C for 1 minute; and a final extension step at 72^{0}C for 7 minutes.

Amplification products were separated by vertical electrophoresis through 8% acrylamide-bisacrylamide (39:1) nondenaturing gels and visualized by silver staining. Allele size was determined using standard markers, ϕX174 (HaeIII cut) and λ-phage (PstI cut). Moreover, some alleles, which were subsequently used as references, were confirmed by DNA sequencing.

Genetic diversity in each ethnic group was estimated as $1-\Sigma p_i^2$, where p_i is the frequency of the ith allele at the locus. Other diversity measures and pairwise $(\delta\mu)^2$ distances between populations were calculated using Microsat (available at http://lotka.stanford.edu/microsat.html). A UPGMA tree was reconstructed from the $(\delta\mu)^2$ distance using the PHYLIP 3.5c package (Felsenstein 1993).

Using primers described in Poggi et al. (1990), exons 2–9 (1059 bp total) and introns 3 and 6 (276 bp total) of the G6PD gene were sequenced in 48 male individuals, according to standard protocols for the direct sequencing of PCR products (Nickerson et al. 1998). Six Akha, 5 Hmong, 6 Karen, 5 Lahu, 5 Lisu, 6 Yao, and 5 Kensiu (Negritos, sometimes referred to as 'Sakai' in Thai or Orang Asli in Malay) from Thailand; and 5 Muong and 5 Kinh (Vietnamese) from Vietnam were included. G6PD deficient individuals were identified in the Thai populations by a biochemical test performed at the Chiang Mai University Medical School (Human Genetics Unit, 1989). All deficient males identified in this manner were included in the sequencing study.

Results

Table 1 displays allele frequencies for the three Y-chromosomal microsatellites in each of four populations. The Karen appear to have greater diversity both in terms of the number of alleles and the evenness of their allele frequency distributions relative to the other three populations. This greater diversity is quantified in Table 2, which presents several measures of genetic variation. The Karen are more variable for every measure calculated. Few of these differences may be significant with the small number of samples and loci currently examined, but the consistency of the various measures — not all of which are independent, however — is reassuring. Importantly, the Hmong data were collected from four distinct villages (which should inflate genetic diversity), while only a single village was sampled for the other three populations. In other words, a single Karen village contains more Y-chromosomal diversity than combined estimates from 4 separate Hmong villages.

Table 3 presents haplotype data for two neighboring Hmong (Miao) groups. Strikingly, only two of eleven 3-locus haplotypes are shared between these groups. In both cases, only a single individual shares the haplotypes. Neither of the two most frequent haplotypes in each group is found in the other population. Though sample sizes are too small to permit a complete analysis of the Y-chromosomal substructure of these two subpopulations, the observed distribution is clearly consistent with low levels of male gene flow between these patrilocal, Hmong groups.

As a way of assessing the phylogenetic information in this data set of 3 Y-chromosomal microsatellite loci, a UPGMA tree was constructed from $(\delta\mu)^2$ distances among the 4 populations for which data were available (Figure 1). The tree is consistent with the known linguistic affinities among the populations. The three Sino-Tibetan speaking populations (Karen, Akha, and Lisu) form a cluster, and the two Burmese-Lolo subfamily members (Akha and Lisu) group together within it. The Hmong, members of a distinct linguistic family (Miao-Yao or Hmong-Mien), comprise an outgroup to the Sino-Tibetan speaking populations.

Finally, Table 4 presents the results of a biochemical test for G6PD deficiency alleles in males. The test is unable to detect heterozygous females, so — apart from noting one homozygous, deficient female in the Hmong population—results for males only are presented. As described in Methods, 48 males were sequenced for the entire coding sequence of the gene and two short introns. All five deficient Karen individuals had the previously described G6PD Mahidol allele (Gly163Ser) (Vulliamy et al. 1989), while the two deficient Yao individuals had the previously described G6PD Quing Yuan allele (Gly131Val) (Chiu et al. 1993). No other DNA variation was detected elsewhere in the gene in any of the samples.

Discussion

Although Y-chromosomal data alone are unable to discriminate between sex-specific differences in migration rates and differences in total or male effective population sizes, the results presented here are consistent with greater male gene flow in the uxorilocal Karen than in neighboring patrilocal populations. Higher male gene flow is expected to increase Y-chromosomal variation within populations, as has been observed for the Karen. Additionally, the extent of Y chromosomal haplotype sharing between adjacent Hmong villages is quite low, and is inconsistent with extensive male gene flow. Without data from mtDNA, the autosomes and, ideally, the X chromosome, it is difficult to attribute greater diversity to migration rate differences alone. It is possible, for example, that the total effective population size of the Karen is larger than for the Hmong, Akha, or Lisu. One might expect recently established communities to harbor greater genetic diversity, if its members have arrived from diverse communities. This seems unlikely to be the case, as the Karen community we sampled is the oldest of any of the settlements included in our study. Because an overall larger N_e should affect all parts of the genome nearly equally, the extent of these contributions to genetic diversity will become clear when we have finished collecting data from mitochondrial DNA and the autosomes.

Some uncertainties about the precise cultural practices of the Hmong, Akha, and Lisu also remain, and it may prove difficult to distinguish some of these from migration rate differences. For example, although polygyny is absent or rare in the Karen, it may occur in some or most Hmong communities. This would reduce the male effective population size among the Hmong, which could reduce the level of Y-chromosomal diversity, mimicking some of the patterns produced by differing male vs. female migration rates. Additionally, the current rules of postmarital residence in the Lisu remain somewhat unclear. Though presumably patrilocal today, there are indications that uxorilocality was practiced during the very recent past. Many of the minority communities were established in Thailand only recently, following modern migrations from Southern China. Thus, analyses may be complicated by these more recent migratory processes, and it is possible that equilibrium has not been achieved in any of them. Precise demographic information for each of the sampled populations will undoubtedly be important in estimating the relative contributions of effective population size and migration rate differences to the genetic measures of population substructure; although most of these questions should yield to more thorough genetic analyses with data from throughout the genome.

Acknowledgments

We wish to thank all of the DNA donors for their samples, time, and interest in this project. We thank Nipa Lajaroj and Sarapee Sila of the Hill Tribes Research Institute in Chiang Mai for their insights, discussions, and detailed ethnographic and demographic knowledge. In addition, both were instrumental in organizing the field research and DNA sample collection. Paiboon Duangchan of Srinakarinviroj University provided valued linguistic and ethnographic data on the Kensiu speaking population of Southern Thailand and Northern Malaysia. We are also grateful to Torpong Sanguansermsri and Kanlaya Payu of the Chiang Mai University Medical School for performing the biochemical test of G6PD enzyme activity (deficiency). Thavorn Vajrabhaya graciously provided some of the samples from Hmong populations. DNA samples from Vietnam were generously provided by Vu Trieu Anh of the Hanoi Medical College. Laboratory and field expenses were supported with funds from the Department of Biology, Chiang Mai University and the L.S.B. Leakey Foundation.

References

Cann, R.L., Stoneking, M. and Wilson, A.C. 1987. Mitochondrial DNA and human evolution. *Nature* 325:31-36.

Carvajal-Carmona, L., Soto, I., Pineda, N., Ortiz-Barrientos, D., Duque, C., Ospina-Duque, J., McCarthy, M., et al. 2000. Strong Amerind/white sex bias and a possible Sephardic contribution among the founders of a population in northwest Colombia. *Am. J. Hum. Genet.* 67:1287-1295.

Cavalli-Sforza, L.L., Menozzi, P. and Piazza, A. 1993. Demic expansions and human evolution. Science 259:639-646.

Chiu, D., Zuo, L., Chao, L., Chen, E., Louie, E., Lubin, B., Liu, T. and Du, C. 1993 Molecular characterization of glucose-6-phosphate deficiency (G6PD) in patients of Chinese descent and identification of new base substitutions in the human G6PD gene. *Blood* 81:2150-2154.

Collins, A., Lonjou, C. and Morton, N.E. 1999. Genetic epidemiology of single-nucleotide polymorphisms. *Proc. Natl. Acad. Sci. USA* 96:15173-15177.

Felsenstein, J. 1993. PHYLIP Phylogeny Inference Package. Seattle, WA: University of Washington.

Human Genetics Unit. 1989. Department of Pediatrics, Faculty of Medicine, Chiang Mai University. *Thalassemia: A Handbook for Laboratory Test.*

Jorde, L.B. and Bamshad, M. 2000. Questioning evidence for recombination in human mitochondrial DNA. *Science* 288:1931.

Jorde, L.B., Watkins, W.S., Bamshad, M.J., Dixon, M.E., Ricker, C.E., Seielstad, M.T., and Batzer M.A. 2000. The distribution of human genetic diversity: a comparison of mitochondrial, autosomal, and Y-chromosome data. *Am. J. Hum. Genet.* 66:979-988.

Kimura, M. 1962. On the probability of fixation of mutant genes in a population. *Genetics* 47:713-719.

Kivisild, T. and Villems, R. 2000. Questioning evidence for recombination in human mitochondrial DNA. *Science* 288:1931.

Kumar, S., Hedrick, P., Dowling, T. and Stoneking, M. 2000. Questioning evidence for recombination in human mitochondrial DNA. *Science* 288:1931.

Mesa, N., Mondragon, M., Soto, I., Parra, M., Duque, C., Ortiz-Barrientos, D., Garcia, L., et al. 2000. Autosomal, mtDNA and Y chromosome diversity in Amerinds: pre and post-Columbian patterns of gene flow in South America. *Am. J. Hum. Genet.* 67:1277-1286.

Nickerson, D., Taylor, S., Weiss, K., et al. 1998. DNA sequence diversity in a 9.7-kb region of the human lipoprotein lipase gene. *Nat. Genet.* 19:233-240.

Nunney, L. 1991. The influence of age structure and fecundity on effective population size. *Proc. R. Soc. Lond. B* 246:71-76.

Nunney, L. 2000. Social structures and effective population sizes. Paper presented at Social Systems and Population Genetics. La Sage, September 6-9.

Parsons, T.J. and Irwin, J.A. 2000. Questioning evidence for recombination in human mitochondrial DNA. *Science* 288:1931.

Perez-Lezaun, A., Calafell, F., Comas, D., Mateu ,E., Bosch, E., Martinez-Arias, R., Clarimon, J., et al. 1999. Sex-specific migration patterns in Central Asian populations, revealed by analysis of Y-chromosome short tandem repeats and mtDNA. *Am. J. Hum. Genet.* 65:208-219.

Poggi, V., Town, M., Foulkes, N.S. and Luzzatto, L. 1990. Identification of a single base change in a new human mutant glucose-6-phosphate dehydrogenase gene by polymerase-chain-reaction amplification of the entire coding region from genomic DNA. *Biochem. J.* 271:157-160.

Poloni, E.S., Semino, O., Passarino, G., Santachiara-Benerecetti, A.S., Dupanloup, I., Langaney, A. and Excoffier, L. 1997. Human genetic affinities for Y-chromosome P49a,f/TaqI haplotypes show strong correspondence with linguistics. *Am. J. Hum. Genet.* 61:1015-1035.

Ruwende, C., Khoo, S., Snow, R. et al. 1995. Natural selection of hemi- and heterozygotes for G6PD deficiency in Africa by resistance to severe malaria. *Nature* 376:246-249.

Seielstad, M., Bekele, E., Ibrahim, M., Touré, A., and Traoré, M. 1999. A view of modern human origins from Y chromosome microsatellite variation. *Genome Res.* 9:558-567.

Seielstad, M.T., Minch, E. and Cavalli-Sforza, L.L. 1998. Genetic evidence for a higher female migration rate in humans. *Nat. Genet.* 20:278-280.

Seielstad, M. 2000. Asymmetries in the maternal and paternal genetic histories of Colombian populations. *Am. J. Hum. Genet.* 67:1062-1066.

Underhill, P.A., Shen, P., Lin, A.A., Jin, L., Passarino, G., Yang, W.H., Kauffman, E., et al. 2000. The architecture of Y-chromosome biallelic haplotype diversity: an emerging portrait of mankind. *Nat. Genet.* 26:358-361.

Vigilant, L., Stoneking, M., Harpending, H., Hawkes, K. and Wilson, A.C. 1991. African populations and the evolution of human mitochondrial DNA. *Science* 253:1503-1507.

Vulliamy, T., Wanachiwanwin, W., Mason, P. and Luzzatto, L. 1989. G6PD mahidol, a common deficient variant in South East Asia is caused by a (163) glycine-serine mutation. *Nuc. Acids Res.* 17:5868.

Table 1. Allele frequencies for four northern Thai populations typed at three Y chromosome microsatellite loci

STR locus	Allele (bp)	Karen	Hmong	Akha	Lisu
		n = 21	n = 29	n = 15	n = 12
DYS19					
	190	0.143	0	0.067	0
	194	0.619	0.276	0.667	0.417
	198	0.238	0.448	0	0.333
	202	0	0.069	0.267	0.249
	206	0	0.207	0	0
DYS389					
	249	0.571	0.828	1	0.917
	253	0.19	0.172	0	0
	257	0.19	0	0	0.083
	261	0.048	0	0	0
DYS393					
	120	0.476	0.345	0	0.75
	124	0.286	0.276	0.733	0
	128	0.19	0.034	0.267	0.167
	132	0.048	0.345	0	0.083

Table 2. Diversity statistics for 3 Y chromosome microsatellite loci in four populations of northern Thailand

Population (N)	Average Hetero-zygosity	Total Hetero-zygosity	Total Variance	Average Alleles	Total Alleles	Average Range
Karen (21)	0.597	0.866	173.489	3.67	11	2.667
Hmong (29)	0.549	0.850	162.176	3.33	10	2.333
Akha (15)	0.290	0.763	162.143	2.00	6	1.333
Lisu (12)	0.287	0.762	171.856	2.67	8	1.333
Average	0.431	0.810	167.416	2.92	8.75	1.917

Table 3. Y chromosome haplotype counts for two Hmong groups in northern Thailand. In the order of DYS19-DYS389-DYS393

Haplotype	Group 1	Group 2
194—249—120	3	0
194—249—124	1	3
194—253—124	1	0
198—249—124	1	1
198—249—132	0	7
198—253—120	1	0
198—253—132	3	0
202—249—120	0	1
202—249—128	0	1
206—249—120	0	5
206—249—124	0	1
Total	10	19

Table 4. Frequency (in males) and identity of G6PD deficiency alleles found in six populations in Thailand (frequencies were determined by biochemical test; allele identity was determined by DNA sequencing)

Population	Frequency	Mutation
Karen	0.21 (24)	Gly163Ser (G→A; Mahidol)
Yao	0.13 (15)	Gly131Val (G→T; Quing Yuan)
Hmong	0.00 (9)	(one deficient female)
Akha	0.00 (15)	
Lahu	0.00 (15)	
Lisu	0.00 (14)	

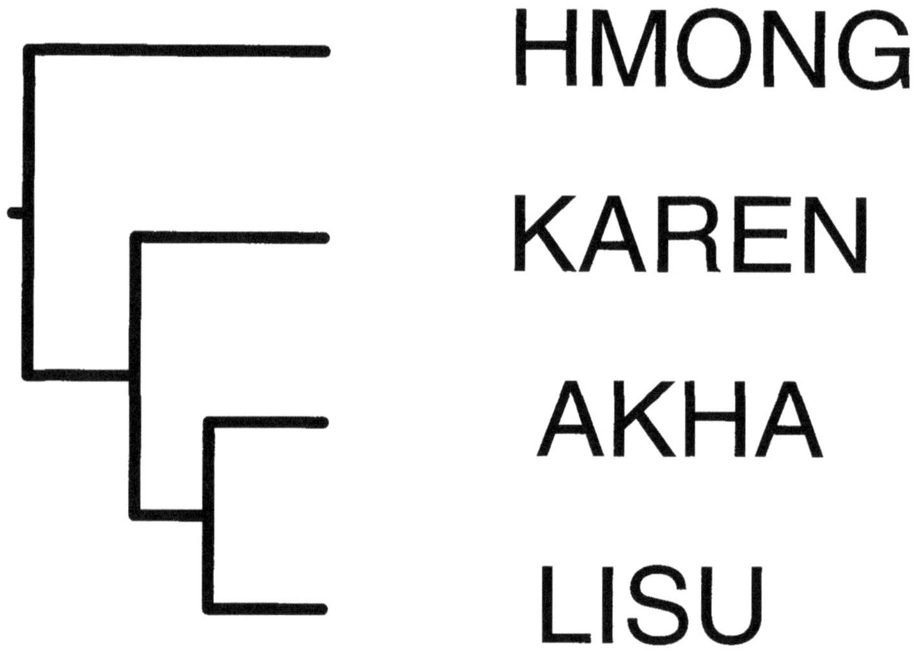

Figure 1. UPGMA tree constructed from 3-locus Y-chromosomal microsatellite distance matrix $(\delta\mu)^2$ among four populations residing in Northern Thailand.

GENETIC RELATIONSHIPS AMONG SIXTEEN ETHNIC GROUPS FROM MALAYSIA AND SOUTHEAST ASIA

S.G. TAN

Department of Biology, Universiti Putra Malaysia

Abstract

Gene frequencies for five polymorphic protein-coding loci namely phosphoglucomutase I, adenosine deaminase, 6-phosphogluconate dehydrogenase, haptoglobin and transferrin were used to calculate genetic distances between pairs of populations. The genetic relationships among eight ethnic groups from Malaysia and eight ethnic groups from other parts of Southeast Asia were obtained by clustering based on their genetic distances using UPGMA. Two major clusters were obtained, one containing three non-Mongoloid ethnic groups, Indians, Senoi and Aetas and the other containing 13 Mongoloid ethnic groups, Kadazan, Malays, Chinese, Bataks, Land Dayaks, Iban, Visayan, Tagalog, Paiwan, Aboriginal Malays, Ifugao, Atayal and Bunun.

Introduction

Malaysia is a multi-racial, multi-lingual, multi-religious and multi-cultural country situated at the crossroads of Southeast Asia. It has a total population of about 20.6 million people of which 16.5 million can be found in Peninsular Malaysia on the Southeast Asian mainland and 1.9 and 2.2 million, respectively in the states of Sarawak and Sabah across the South China Sea on the island of Borneo. The multi-ethnic character of the Malaysian population had arisen over the last 150 years and generally Malaysia's ethnic groups fall into two main categories; those categorized as being indigenous to the region (Bumiputeras or sons of the soil in the Malay language) make up 59% of the total population and those whose ancestors migrated from elsewhere mainly from China (32%) and India (8%) but with contributions from elsewhere such as Arabia and Europe (Anon. 1998).

The indigenous groups themselves are heterogeneous and fall into three main categories namely Peninsular Malaysia Aborigines (Orang Asli), Malays and the indigenous peoples of Sabah and Sarawak. The Orang Asli are the smallest and oldest group in the country and are currently found in small numbers (92,529) and in scattered groups in the peninsula (Nicholas 1996). They consist of three major groups Negrito (2,972), Senoi (49,440) and Aboriginal or Proto-Malays (40,117) each with their own subgroups. Baer (1999) had summarized and evaluated the published biomedical and genetic data available on the Orang Asli of Malaysia.

The Muslim Malays form the main ethnic group in the peninsula but substantial numbers are also found in the two Borneo states (Anon. 1998). The Malays of the east coast of the peninsula and Sabah and Sarawak have long settled in Malaysia whereas those on the west coast of the peninsula are mainly descended from those who crossed the narrow Straits of Malacca from Sumatra and settled in considerable numbers in the late nineteenth and early twentieth centuries. Other Muslim groups, descendants of migrants from various other Indonesian islands, who are now considered as Malays include Javanese especially found in the west coast of Johore, Selangor and Perak states, Banjarese, Boyanese, Bugis and the Minangkabau, who are unusual in that this Sumatra originating group follows the matriarchal social system.

In Sarawak, the Iban (Sea Dayak) form the largest ethnic group comprising 30% of the state's population followed by the Chinese (28%), Malay (20%), Bidayuh or Land Dayak (8%), Melanau (5%) and other indigenous groups such as the Kenyah, Kayan, Ukit, Penan, Sekapan, Lahanan, Lun Bawang, Kelabit, Berawan and Punan Bah who together constitute 5% of the population (Phoa 1996). In Sabah, the 39 indigenous groups together make up 83% of the state's population with the Chinese making up 16%. Of the indigenous groups, the Kadazan or Kadasan-Dusun group is the largest making up 24% of Sabah's population followed by Bajau (8%), Murut (3%), Paitan (2.6%) and Suluk (1.3%). Other smaller indigenous groups in Sabah include the Bonggi, Illanun, Bengkahak or Mangkaak and Tidung (Lasimbang 1996).

With such an assortment of ethnic groups it is not surprising that this country had attracted the attention of human population geneticists since the 1960s. The most productive group was from the University of California San Francisco's International Centre for Medical Research headed by the late Dr. Luan Eng Lie-Injo who was based at the Institute for Medical Research,

Kuala Lumpur, Malaysia. Besides studying thalassaemia and abnormal hemoglobin they also studied blood groups and protein electrophoretic variations from erythrocytes and serum. Their results were published in numerous papers in many well known journals such as Human Genetics, Human Heredity and the American Journal of Human Genetics. I was fortunate to have been able to collaborate with this group from 1976 to 1980 and compiled most of their population genetics results in three summary papers (Tan 1978a, 1978b, 1979). Subsequent to that some studies using cerumen (Norakmal and Tan 1979), human placenta (Teng et al. 1978a, 1978b; Teng and Tan 1979) and saliva (Tan and Teng 1978, 1979; Noraini et al. 1980) as sources of genetic markers and isoelectric focusing subtyping of serum group specific component (Gc) (Tan et al. 1981) and transferrin (Tan et al. 1982) were done. While there is a rich data base on the major ethnic groups of Malaysia for allozyme loci, there is much less data on DNA based markers (Ballinger et al. 1992; Parra et al. 1999a, 1999b). Hence, it is imperative that Malaysia's rich human biodiversity be studied using DNA based genetic markers such as microsatellites before ethnic boundaries are further blurred through increased inter-ethnic marriages as a result of rapid urbanisation and advances in communications and transportation. As it is in Sabah state now Sino-Natives make up about 1.2 % of the population (Lasimbang 1996).

Here, an attempt is made to find the genetic relationships among Malaysians of various ancestries, Malay, Chinese, Indian, Senoi, Aboriginal Malay from the Peninsular Malaysia, Kadazan from Sabah, Land Dayak and Iban from Sarawak with other Southeast Asian ethnic groups namely the Atyal, Bunun and Paiwan, three indigenous groups from Taiwan, Aetas (Negrito), Tagalog, Visayan and Ifugao from the Philippines and Batak from Sumatra, Indonesia based on the published data for five polymorphic allozyme loci.

Materials and Methods

A list of gene frequencies for five protein level polymorphic loci, phosphoglucomutase I, adenosine deaminase, 6-phosphogluconate dehydrogenase, haptoglobin and transferrin, which had been typed and reported by various authors for the 16 ethnic groups found in Southeast Asia previously mentioned were obtained from the tabulation done by Tan (1979). These allele frequencies were used to calculate the genetic distance (Nei 1978) between populations by the BIOSYS-1 software package (Swofford and Selander 1989). These distances are presented in Table 1 and an Unweighted Pair Group Method with Arithmetic Averages (UPGMA) (Sneath and Sokal 1973) dendrogram was generated to determine the genetic relationships among the 16 ethnic groups.

Results and Discussion

The dendrogram showing the genetic relationships among 16 ethnic groups found in Southeast Asia is presented in Figure 1. Two major clusters can be clearly differentiated. One major cluster contains the Indians, Senoi and Aetas which are the non-Mongoloid ethnic groups. Malaysian Indians are mostly the descendants of southern Indian immigrants who are Dravidians, "pure" Senoi are Veddoids but present day Senoi are probably the descendants of past interbreeding between "pure" Senoi and Negrito, some of the Mongoloid ethnic groups and perhaps even Indians whereas Negrito, a pygmy ethnic group was the first of the present ethnic groups to settle in Southeast Asia (Provencher 1975). The other major cluster contains the Mongoloid ethnic groups of Southeast Asia. Malaysians of Chinese ancestry are the descendants of immigrants from southern China who belong to the Southern or Palaeo- Mongoloid group of ethnic groups just as the Kadazan, Malays, Bataks, Land Dayaks, Iban, Visayan, Paiwan, Tagalog, Aboriginal Malays, Ifugao, Atyal and Bunun.

Among the Mongoloid ethnic groups, the Kadazan forms a cluster, the Malays, Chinese and Bataks of Sumatra cluster together, the Land Dayaks and Iban of Sarawak, the Visayan and Tagalog of the Philippines and the Paiwan from Taiwan form another cluster while the Aboriginal Malays of the Peninsular Malaysia, Ifugao from the Philippines and Atayal and Bunun of Taiwan formed a third cluster. Since this clustering is only based on five polymorphic loci it may change if more loci had been available. However, it did clearly separate the Mongoloid from the non-Mongoloid ethnic groups. The fact that the Kadazans cluster alone is interesting since their ethnic

Table 1. Genetic distances (Nei, 1978) between 16 ethnic groups found in Southeast Asia

	Kadazan	Malays	Senoi	AbgMal	Chinese	Indians	Dayaks	Iban	Atyal	Bunun	Paiwan	Aetas	Tagalog	Visayan	Ifugao	Bataks
Kadazan	0.0000	-	-	-	-	-	-	-	-	-	-	-	-	-	-	-
Malays	0.0197	0.0000	-	-	-	-	-	-	-	-	-	-	-	-	-	-
Senoi	0.0198	0.0126	0.0000	-	-	-	-	-	-	-	-	-	-	-	-	-
AbgMal	0.0079	0.0084	0.0248	0.0000	-	-	-	-	-	-	-	-	-	-	-	-
Chinese	0.0116	0.0021	0.0068	0.0067	0.0000	-	-	-	-	-	-	-	-	-	-	-
Indians	0.0443	0.0091	0.0122	0.0343	0.0131	0.0000	-	-	-	-	-	-	-	-	-	-
Dayaks	0.0059	0.0082	0.0117	0.0083	0.0051	0.0219	0.0000	-	-	-	-	-	-	-	-	-
Iban	0.0041	0.0063	0.0132	0.0027	0.0028	0.0253	0.0021	0.0000	-	-	-	-	-	-	-	-
Atyal	0.0079	0.0051	0.0142	0.0019	0.0025	0.0242	0.0058	0.0016	0.0000	-	-	-	-	-	-	-
Bunun	0.0085	0.0136	0.0205	0.0041	0.0084	0.0385	0.0111	0.0051	0.0023	0.0000	-	-	-	-	-	-
Paiwan	0.0049	0.0096	0.0068	0.0087	0.0039	0.0231	0.0025	0.0025	0.0039	0.0062	0.0000	-	-	-	-	-
Aetas	0.0356	0.0243	0.0049	0.0458	0.0193	0.0126	0.0204	0.0280	0.0318	0.0422	0.0179	0.0000	-	-	-	-
Tagalog	0.0073	0.0059	0.0066	0.0071	0.0019	0.0185	0.0027	0.0020	0.0023	0.0055	0.0006	0.0182	0.0000	-	-	-
Visayan	0.0057	0.0062	0.0068	0.0062	0.0013	0.0205	0.0038	0.0015	0.0025	0.0065	0.0017	0.0202	0.0013	0.0000	-	-
Ifugao	0.0062	0.0073	0.0212	0.0006	0.0056	0.0304	0.0048	0.0015	0.0015	0.0043	0.0059	0.0389	0.0048	0.0049	0.0000	-
Bataks	0.0113	0.0017	0.0098	0.0050	0.0004	0.0148	0.0044	0.0020	0.0022	0.0079	0.0046	0.0231	0.0023	0.0019	0.0041	0.0000

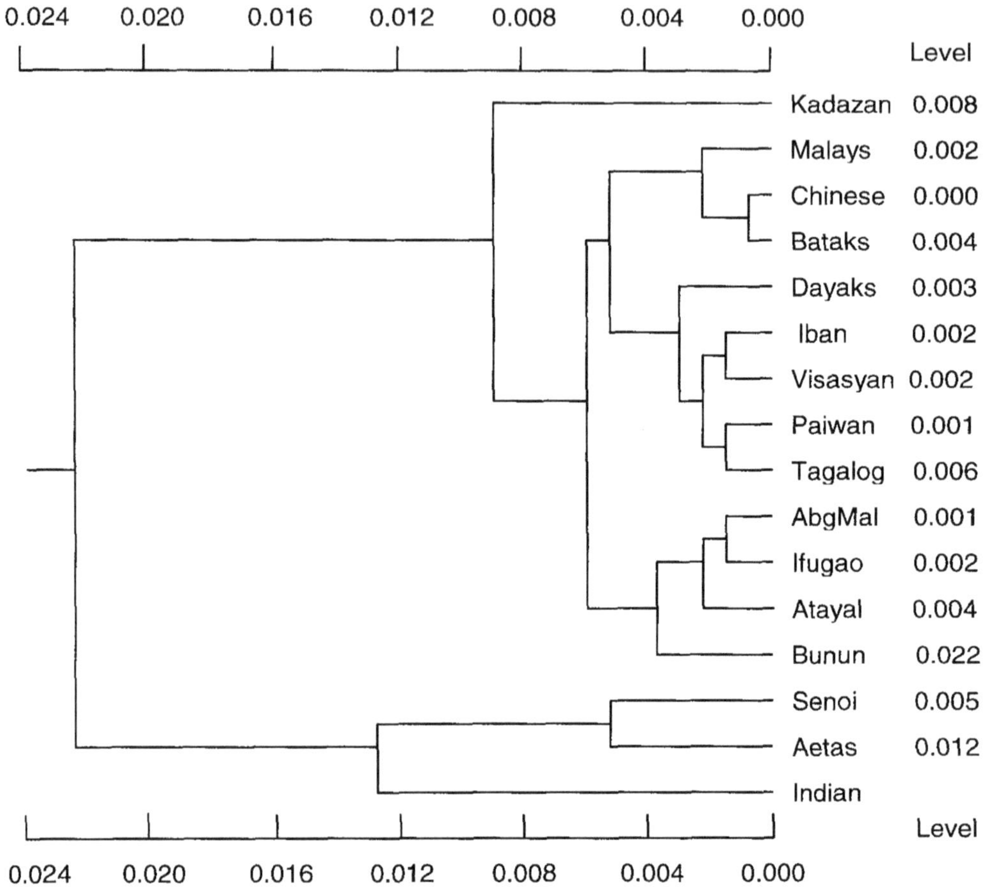

Figure 1. UPPGMA dendrogram of 16 ethnic groups found in Southeast Asia based on the genetic distance of Nei (1978).

affinities are uncertain (Tan et al. 1979). Some anthropologists for example Williams (1965) hold the view that they are physically, linguistically and culturally closer to a number of native groups in the Philippines and Taiwan than to the native peoples of neighbouring Sarawak and Kalimantan (Indonesian Borneo). These anthropologists think that the Kadazans are descended from a second wave of migration of Indo-Malayan agriculturalists, who 3000 years ago, moved from south China and north Vietnam through Taiwan and the Philippines to northern Borneo and Celebes (Sulawesi). The natives of Sarawak and Kalimantan, however, are believed to have originated from a first wave of migrations slightly earlier, who spread across mainland Southeast Asia into northern India and down the Malay Peninsula into Sumatra, Java and south Borneo and from there into Sarawak (Williams 1965; Harrison 1964). Other anthropologists think that the Kadazans migrated to northern Borneo from the Malay Peninsula only a few centuries ago (Provencher 1975).

With regards to the migration of humans to Southeast Asia, our studies on swamp water buffalo populations from Malaysia (Peninsular, Sabah and Sarawak) Thailand, the Philippines (Mindanao), Indonesia (Sumatra, Java and Celebes) using 53 protein-coding loci (Barker et al. 1997a), 21 microsatellite loci (Barker et al. 1997b) and 303 bp mitochondrial DNA cytochrome b and 158 bp D-loop sequences (Lau et al. 1998) are relevant since these domestic animals were brought to Southeast Asia by human rice growing migrants. The swamp buffalo was domesticated in China about 7000 years ago (Chen and Li 1989), probably following the development of the rice culture in the lower Yangtze region about 8000 years ago (Bellwood 1992). The spread of this rice culture together with the swamp buffalo from southern China to mainland Southeast Asia dates from some 5000 years ago and to Sumatra and Java about 3000–4000 years ago. The results of our molecular population genetics study of buffaloes support the following hypothesis for the migration of the swamp buffalo (and hence humans) into Southeast Asia (Lau et al. 1998). After domestication in China, the domesticated swamp buffalo spread with rice cultivation through (a) Taiwan, to the Philippines and the eastern islands of Borneo and Celebes, and (b) south through mainland Southeast Asia (likely interbreeding with wild buffalo) to Peninsular Malaysia and onto the western islands of Indonesia (Sumatra and Java). Hence it will be interesting to see whether similar studies on the human populations of Southeast Asia using DNA level markers will support this scenario of human migration into this region from the centre of origin of the Southern Mongoloids in the northern Vietnam or southern China region (Ballinger et al. 1992). Therefore, with her geographical location at the centre of Southeast Asia and her numerous ethnic groups, Malaysia should play a prominent role in this important endeavour.

Acknowledgement

I thank my postgraduate students, Ms. Pung Chai Chin, Mr. V.J. Kumar and Mr. Abdul Latif for assistance in the preparation of this paper.

References

Anon. 1998. *Information Malaysia 1998 Yearbook*. Kuala Lumpur:Berita Publishing.

Baer, A. 1999. *Health, Disease and Survival: A Biomedical and Genetic Analysis of the Orang Asli of Malaysia*. Subang Jaya:Center for Orang Asli Concerns.

Ballinger, S.W., Schurr, T.G., Torroni, A., Gan, Y.Y., Hodge, J.A., Hassan, K., Chen, K.H., et al. 1992. Southeast Asian mitochondrial DNA analysis reveals genetic continuity of ancient Mongoloid migrations. *Genetics* 130:139-152.

Barker, J.S.F., Tan, S.G., Selvaraj, O.S., and Mukherjee, T.K. 1997a. Genetic variation within and relationships among populations of Asian water buffalo (*Bubalus bubalis*). *Animal Genet.* 28:1-13.

Barker, J.S.F., Moore, S.S., Hetzel, D.J.S., Evans, D., Tan, S.G., and Byrne, K. 1997b. Genetic diversity of Asian water buffalo (*Bubalus bubalis*): microsatellite variation and a comparison with protein-coding loci. *Animal Genet.* 28:103-115.

Bellwood, P. 1992. Southeast Asia before history. IN: Tarling, N. (ed.), *The Cambridge History of Southeast Asia Vol. 1*. Cambridge: Cambridge University Press, pp. 55-136.

Chen, Y.C. and Li, X.H. 1989. New evidence of the origin and domestication of the Chinese swamp buffalo (*Bubalus bubalis*). *Buffalo J.* 1:51-55.

Harrison, T. 1964. The people of north and west Borneo. IN: Wang, G. (ed.), *Malaysia*. Singapore: Donald Moore, pp. 163-178.

Lasimbang, J. 1996. The indigenous peoples of Sabah. IN: Nicholas, C. and Singh, A. (eds.), *Indigenous Peoples of Asia*. Bangkok:Asia Indigenous Peoples Pact, pp. 177-195.

Lau, C.H., Drinkwater, R.D., Yusoff, K., Tan, S.G., Hetzel, D.J.S., and Barker, J.S.F. 1998. Genetic diversity of Asian water buffalo (*Bubalus bubalis*):

Mitochondrial DNA D-loop and cytochrome b sequence variation. *Animal Genet.* 29:253-264.

Nei, M. 1978. Estimation of average heterozygosity and genetic distance from a small number of individuals. *Genetics* 89:583-590.

Nicholas, C. 1996. The Orang Asli of Peninsular Malaysia. IN: Nicholas, C. and Singh, A. (eds.), *Indigenous peoples of Asia.* Bangkok: Asia Indigenous Peoples Pact, pp. 157-176.

Noraini, I., Tan, S.G., Gan, Y.Y., and Teng, Y.S. 1980. Salivary peroxidase, Pm and Ph protein polymorphisms in Malaysians. *Hum. Genet.* 56:205-207.

Norakmal, I. and Tan, S.G. 1979. Cerumen polymorphism in the three major ethnic groups of Malaysia. *Jap. J. Hum. Genet.* 24:119-121.

Parra, E., Saha, N., Soemantri, A.G., McGarvey, S.T., Hundrieser, J., Shriver, M.D., and Deka, R. 1999a. Genetic variation at 9 autosomal microsatellite loci in Asian and Pacific populations. *Hum. Biol.* 71:757-779.

Parra, E., Shriver, M.D., Soemantri, A., McGarvey, S.T., Hundrieser, J., Saha, N., and Deka, R. 1999b. Analysis of five Y-specific microsatellite loci in Asian and Pacific populations. *Am. J. Phys. Anthrop.* 110:1-16.

Phoa, J. 1996. The Dayaks and Orang Ulu of Sarawak. IN: Nicholas, C. and Singh, A. (eds.), *Indigenous Peoples of Asia.* Bangkok:Asia Indigenous Peoples Pact, pp. 197-212.

Provencher, R. 1975. *Mainland Southeast Asia: An Anthropological Perspective.* Pacific Palisades:Goodyear Publishing Company.

Sneath, P.H. and Sokal, R.R. 1973. *Numerical Taxonomy.* San Francisco:Freeman.

Swofford, D.L., and Selander, R.B. 1989. *Biosys-1: A computer program for the analysis of allelic variation in population genetics and biochemical systematics.* Release 1.7. Urbana-Champaign: Illinois Natural History Survey.

Tan, S.G. 1978a. Biochemical genetic markers in the three main races of Peninsular Malaysia and Singapore: A compilation of data. *Pertanika* 1:22-35.

Tan, S.G. 1978b. Genetic markers in the Aborigines of Peninsular Malaysia and the Indigenous races of Sabah, Sarawak and Brunei. Blood groups in the three major races of Malaysia and Singapore: A compilation of data. *Pertanika* 1:86-96.

Tan, S.G. 1979. Genetic relationship between Kadazans and fifteen other Southeast Asian races. *Pertanika* 2:28-33.

Tan, S.G. and Teng, Y.S. 1978. Saliva acid phosphotase and amylase in Senoi and Aboriginal Malays and superoxide dismutase in various racial groups of Peninsular Malaysia. *Jap. J. Hum. Genet.* 23:133-138.

Tan, S.G. and Teng, Y.S. 1979. Saliva acid phosphotases in Malaysians: Report of a new variant. *Hum. Here.* 29:61-63.

Tan, S.G., Teng, Y.S., Ganesan, J., Lau, K.Y., and Lie-Injo, L.E. 1979. Biochemical markers in the Kadazans of Sabah, Malaysia. *Hum. Genet.* 49:349-353.

Tan, S.G., Gan, Y.Y., Asuan, K., and Abdullah, F. 1981. Gc subtyping in Malaysians and Indonesians from North Sumatra. *Hum. Genet.* 59:75-76.

Tan, S.G., Gan, Y.Y., and Asuan, K. 1982. Transferrin C subtyping in Malaysians and Indonesians from North Sumatra. *Hum. Genet.* 60:369-370.

Teng, Y.S. and Tan, S.G. 1979. Acid alpha-glucosidase in Malaysians: Population studies and the occurrence of a new variant. *Hum. Here*. 29:2-4.

Teng, Y.S., Tan, S.G., Ng, T., and Lopez, C.G. 1978a. Red cell glyoxalase I and placental soluble aconitase polymorphisms in the three major ethnic groups of Malaysia. *Jap. J. Hum. Genet*. 23:211-215.

Teng, Y.S., Tan, S.G., Lopez, C.G., Ng, T., and Lie-Injo, L.E. 1978b. Genetic markers in Malaysians: Variants of soluble and mitochondrial glutamic oxaloacetic transaminase and salivary and pancreatic amylase, phosphoglucomutase III and salivary esterase polymorphisms. *Hum. Genet.* 41:347-354.

Williams, T.R. 1965. *The Dusun: A North Borneo Society*. New York:Holt, Rinehart and Winston.

Part III: The Peopling of East Asia

CHINESE HUMAN GENOME DIVERSITY PROJECT: A SYNOPSIS

JIAYOU CHU

Institute of Medical Biology, Chinese Academy of Medical Sciences

The majority of Chinese consisting of the Han nationality (93.3%) and 55 official minority nationalities (6.7%) are found predominantly in the peripheral regions, most of which have their own languages. The number of living languages listed for China is 205 (Grimes 1996). Despite the fact that extensive variations among Han Chinese populations and minority populations in China have been observed (Zhao, et al. 1986; Zhao and Lee 1989; Weng et al. 1989; Zhang, H 1988; Zhang, ZB 1988; Etler 1992); such populations are usually underrepresented in genetic studies of worldwide populations (Bowcock et al. 1994; Deka et al. 1995; Jorede et al. 1995). The significance of an extensive study of Chinese populations is two fold. First, a distinction between northern and southern Chinese populations (Han and minority alike) has been observed in the analyses of genetic markers (Zhao et al. 1986; Zhao and Lee 1989; Weng et al 1989) as well as somatometric and nonmetric features (Zhang, H 1988; Zhang, ZB 1988; Etler 1992). Most authors attribute such distinction simply to the presence of geographic barriers (Zhao et al. 1986; Zhao and Lee 1989; Weng et al. 1989; Zhang, H 1988; Zhang, ZB 1988; Etler 1992). While it is true that a geographic barrier maintains genetic difference if there is any, it is irrelevant to a more interesting question, which is, whether southern and northern populations are descendents of the same population or, alternatively whether, they have arrived in China at different times. Furthermore, an understanding of origins of the populations in East Asia may shed light on the peopling of the Siberia and Americas. Secondly, the human fossil remains recovered in China have also attracted attention. The regional as well as the temporal continuity of fossil records from *H. erectus* to *H. sapiens* in this region (Wang 1986; Brooks and Wood 1990; Li and Etler 1992) has repeatedly challenged the Out-of-Africa hypothesis that suggests a complete replacement of local populations by modern humans originating in Africa -- although the validity of the analyses (Li and Etler 1992) have been questionable (Cann 1996). Genetic evidence became necessary to verify such claims. A systematic genetic study of the Chinese populations using contemporary genetic markers therefore was conducted. This chapter provides a collection of activities made under the auspice of the Chinese Human Genome Diversity Project (CHGDP).

Chinese Human Genome Diversity Project and National Cell Line Repository

Due to a paucity of funding, but the avalability of rich resources for diversity studies, the CHGDP has become the centerpiece of the Chinese Human Genome Project that was inaugurated in 1993. The collection and preservation of the population samples was considered a top priority in the first phase of the CHGDP that extended from 1993 to 1998. Preservation was achieved by converting the B-lymphocytes from venous blood to immortalized cell lines with EB virus transformation, which would generate inexhaustible amounts of DNA for future studies. About 100 samples were collected for each population. Proper informed consent was obtained from each participating subject. The selection of the populations was extensively discussed among the participants of the CHGDP and the sample collection process was closely monitored to ensure the authenticity of the samples in terms of their ethnicity.

So far, the cell lines for 15 populations have been established and they include Dai, Jingpo, Deang, Tibetans both in Tibet and Yunnan, Jino, Yi, Oroqen, Ewenki, Daur, Wa, Naxi, Bai, and Han Chinese from Shangxi and Guangdong. The cell lines from 27 more populations will be established before 2002 and they include Uygur, Kazak, Kirgiz, Xibe, Tajik, Uzbek, Hui, Mongol, Korean, Manchurian, Qiang, Tujia, Miao, Yao, She, Dong, Bouyei, Zhuang, Sui, Lisu, Hezhen, and Han Chinese from Shangdong, Henan, Hunan, Fujian, Sichuan and Zhejiang. Two cell line reservoirs were established: one in the Kunming Institute of Medical Biology, Chinese Academy of Medical Sciences and the other in the Beijing Institute of Genetics, Chinese Academy of Sciences. DNAs will be extracted, stored and distributed for further research. Furthermore, some of the cell lines were shipped to CEPH, from where those samples will be made available to researchers across the world.

As a report of the first phase of the Project, a paper on the genetic relationship of the Chinese populations using microsatellites was published in the Proceedings of National Academy of Sciences (Chu et al. 1998). The next section of this chapter provides a concise summary of this work. This effort was further expanded to the study of Y chromosomal markers, mtDNA, and autosomal markers in the second phase of the Project starting in 1998 (Su et al. 1999, 2000, Qian et al. 2000). Some of the results are also included in Chapters 5 and 9, of this volume.

A Study of the Populations in East Asia Using Microsatelite Markers

Twenty-eight populations which belong to six language families and currently reside in China were studied. Based on their linguistic affiliation, they are they are Sino-Tibetan (Aini, Deang, Jingpo, Lahu, Tibetan, Tujia, Yi, Hui, and four Han Chinese populations from Guangdong, Henan, Beijing, Yunnan), Austroasiatic (Blang, Wa), Daic (Dai, Dong, Li, Yao-Jinxiu), Hmong-Mien (She, Yao-Puno), Altaic (Ewenki, Korean, Manchurian, Uyghur), and Austronesian (Ami, Atayal, Paiwan, Yami). Samples were collected through a coordinated effort of several institutes participating with the CHGDP. Fifteen non-Chinese groups were used as references and they are Japanese, Buryat, Yakut, Cambodian, Karitiana, Mayan, Australian, New Guinea Highlander, Burushaski, Italian, Basque, Europeans derived from CEPH, Pygmies from the Central Africa Republic, Pygmies from Zaire, and Lissongo.

Two Neighbor-joining phylogenies are presented in Figure 1a and 1b (Saitou and Nei 1987). A genetic distance measure, Dc, was used to estimate differentiation between populations (Cavalli-Sforza and Edwards 1967, Nei 1987). Only populations with complete allele frequency data for all microsatellite loci were available was included for analysis. Bootstrap values were obtained based on 500 replications. An African population lineage was used to root the phylogeny based on the result of Bowcock et al. (1994).

In Figure 1, the microsatellites analyzed are D1S484, D2S434, D3S1768, D6S1009, D7S493, D10S537, D12S101, D12S373, D13S126, D15S101, D15S102, D15S230, D16S508, D16S667, D17S1824, D18S465, D19S152, D19S210, D19S414, D19S420, D19S601, D20S100, D20S115, D20S118, D20S171, D20S471, D21S1435, D22S1158, HLIP, and UTSW1523. In this phylogeny, populations and loci were selected to

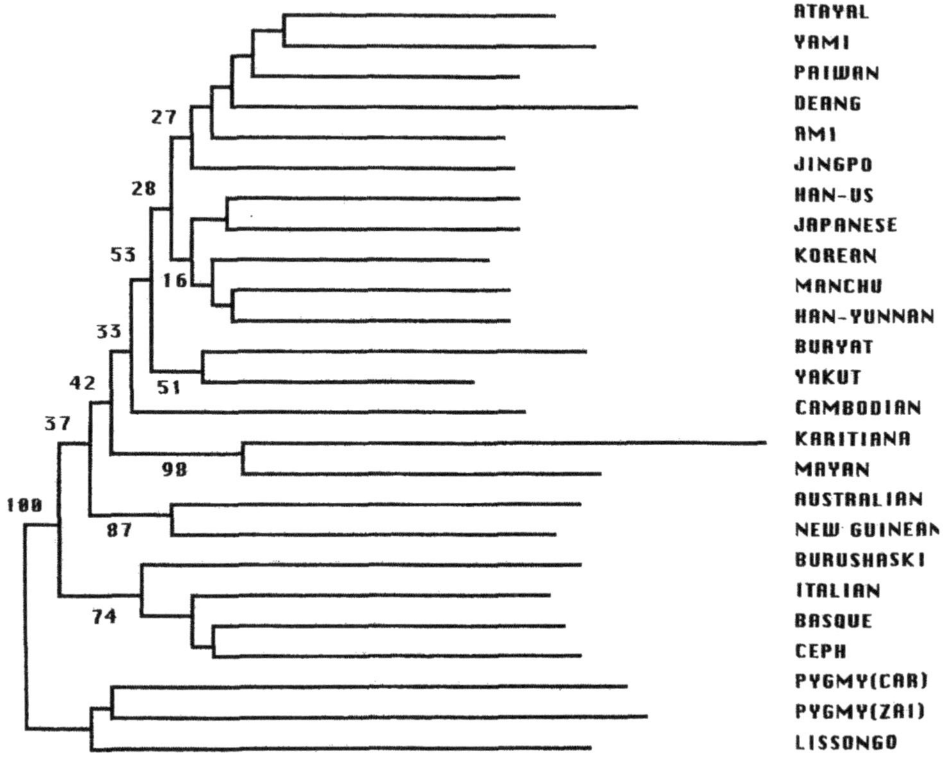

Figure 1. Phylogenies constructed by using the neighbor-joining method based on 30 microsatellites. Numbers on the branches are bootstrap values based on 500 replications.

maximize the number of loci in the analysis. Eight Chinese populations were included in Figure 1. They are Han from Yunnan, Han from Guangdong, Manchurian, Jingpo, Deang, Atayal, Yami, and Paiwan.

In Figure 2, microsatellites analyzed are D1S484, D2S434, D7S493, D10S537, D12S373, D16S667, D17S1824, D19S152, D19S210, D19S414, D19S420, D20S100, D20S115, D20S171, and D21S1435. In this phylogeny, the number of populations for analysis was maximized to achieve the best regional representation while the loci are again selected whenever their allele frequency information is available across those populations. In this analysis, sixteen more Chinese populations were added. They are Uyghur, Han from Beijing, Wa, Jingpo, Tujia, Tibetan, Hui, Ewenki, Yao speaking Puno, Yi, She, Yao from Jinxiu, Han from Henan, Dong, Li, Lahu, Dai, Blang, Aini, and Ami.

The phylogeny based on 30 microsatellites (see Figure 1) revealed a clear distinction between southern and northern Chinese populations, although the number of Chinese populations included in this phylogeny is small. Three northern Chinese populations clustered with the Japanese and Korean as expected. This phylogeny provides validation for our current approach given the fact that the relationship among worldwide populations is identical to that presented in Bowcock et al. (1994) which was derived using a completely different set of markers but some populations analyzed in this study were included in Bowcock et al. (Cambodian, Karitiana, Mayan, Australian, New Guinean, Italian, Zaire Pygmy, Central Republic Pygmy and Lissongo). Populations from East Asia form a distinctive cluster indicating a common ancestry shared among those groups. Taiwanese Aboriginal populations derived from the southern continental population cluster, indicating the probable origin of those populations and probably Polynesians.

The distinction between southern populations and northern populations is noticeable but far less clear in Figure 2 when 16 more Chinese populations were added to the analysis. The number of loci was reduced to 15 due to incomplete data for some of the loci. Again, the populations from East Asia were derived from the same lineage. In Figure 2, two clusters for the northern populations are discernible; Altaic language speaking Buryat, Yakut, Uyghur, and Manchu clustered with the Korean and Japanese, two language isolates but closely related to Altaic. Two Han populations, one from north China and the other from Yunnan also contributed to this cluster (Cluster N1). Another Altiac language speaking population, Ewenki formed

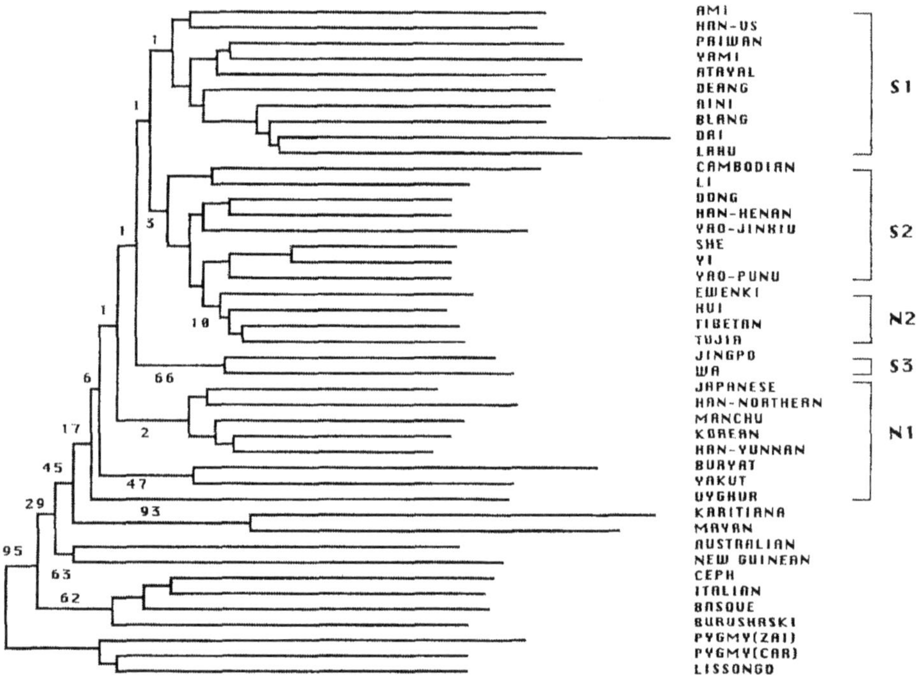

Figure 2. Phylogenies constructed by using the neighbor-joining method based on 15 microsatellites. Numbers on the branches are bootstrap values based on 500 replications.

a cluster (Cluster N2) with Tibetan, Tujia, and Hui, all of which were originally derived from the northern populations, though currently living in the western part of China. Populations of southern origin formed three clusters. In the first southern cluster, Blang, an Austro-Asiatic population, grouped with Deang, Aini, Lahu, and Dai, all sampled from the southwest part of Yunnan. This lineage then clustered with three populations from Taiwan (Paiwan, Atayal, and Yami) probably reflecting the origin of Taiwanese aborigines and thus Polynesian from Southeast Asia. The fourth Taiwanese aboriginal population, Ami, forms a separate cluster with the Han Chinese of southern origin living in the US before they joined the previous cluster to form Cluster S1. The second southern group consists of three Daic populations (Li, Dong, and Yao from Jinxiu) all from Guangxi or Hainan, two Hmong-Mien populations (She and Yao speaking Punu), Cambodian (an Austro-Asiatic population), Yi and Han from Henan (Cluster S2). The second northern lineage (Cluster N2) consisting of mostly western populations derived from this southern group except Ewenki. Jingpo and Wa formed the third southern lineage (Cluster S3).

In this phylogeny, populations in East Asia can be divided into two groups: a northern group consisting of populations in Cluster N1 and a southern group including all southern populations (Clusters S1, S2, S3) and the second cluster of northern origin (Cluster N2). This relationship was not strongly supported by the bootstrap values among major clusters most of which were small. However, a phylogeny with 17 Chinese populations and 8 worldwide populations based on 26 loci presented a very similar topology where the bootstrap value supporting the separation of the first northern cluster with the southern clusters being 13% and that supporting the second northern lineage being 19%.

Validation of the utility of microsatellites in reconstructing the evolutionary history of human populations has been made not only theoretically (Goldstein et al. 1995a; Slatkin 1995; Goldstein et al. 1995b; Deka et al. 1996) although many of such studies used distantly related populations and therefore, the utility of such markers in the study of closely related populations are yet to be explored. The current study reflects, to some extent, a lack of resolution of microsatellites in the reconstruction of the closely related populations, probably due to an insufficient number of loci for a large number of populations studied but less likely due to the insufficient number of samples for each population as demonstrated by Shriver et al. (1997). Small bootstrap values reflect an insufficient amount of information available to resolve the genetic relationship among closely related populations. The presence of strong gene flow among those populations would further compromise the reliability. Nevertheless, it is not our primary intention to reveal the detailed genetic relationship among those closely related populations, but rather we are interested in exploring the major pattern of evolutionary history of the human populations currently residing in East Asia.

In both phylogenies with different loci and populations, populations from East Asia always derived from a single lineage indicating the single origin of those populations. This observation of single origin and its connection with African lineages is clearly consistent with the Out-of-Africa hypothesis. It does not preclude the possibility of an independent origin of modern humans in East Asia, since the current approach can only detect a major genetic contribution from such origin. A phylogeny with very different topological structure would have been expected if an independent Asian origin of modern humans had made a major contribution to the current gene pool in Asian populations. A haplotype-based analysis will probably detect a minor contribution from an independent origin of modern humans in East Asia (Su et al. 1999; Jin and Su 2000).

The current analysis suggests that the southern populations in East Asia may be derived from the populations in Southeast Asia that originally migrated from Africa possibly via mid-Asia, and the northern populations were under strong genetic influences from Altaic populations from the north. But it is unclear how Altaic populations migrated to Northeast Asia. It is possible that ancestral Altaic populations arrived there from the middle of Asia or alternatively they may have originated from East Asia. The haplotype distribution of the Y chromosomal markers showed that the majority of the gene pools in the northern part of East Asia, including the Altaic populations was originally derived from the southern populations although the current Altaic-speaking populations undeniably admixed with later arrivals from mid-Asia and Europe (see Chapter 9, Su and Jin). The topology of the phylogenies based on autosomal microsatellites is consistent with this observation as manifested by the position of Cluster N1 in Figure 2.

The Express Train hypothesis placed Taiwan as the homeland of the Austronesian populations that eventually marched toward Oceania (Diamond 1988, Melton et al. 1995). A recent study based on Y chromosomal markers challenged the validity of this hypothesis and suggested that the migration towards Oceania was independent from that to Taiwan (Su et al. 2000). Therefore, the origin of Taiwanese Aborigines is intriguing. Ami is always clustered with Han Chinese from Guangdong (Han-US), while Yami, Atayal, and Paiwan clustered together and then with southern populations including those speaking Austroasiatic, Daic, and Sino-Tibetan languages. The phylogenies clearly support the notion that Taiwanese Aborigines originally migrated from the southern part of China to Taiwan. However, it is practically difficult to trace the origin of Taiwanese Aborigines since (1) the emigration occurred long before the ethnicity and languages had emerged, and (2) migrations in the southern part of China have been substantial, erasing the genetic footprints of such emigration.

Acknowledgements

I wish to acknowledge my colleagues (Qian Yaping, Yang Zhaoqing, Chu Zhengtao, Yu Jiankun, and Hong Kun Xue) and our collaborators (Jin Li, Li Pu, Xu Jiujin, Zhang Sizhong, Chen Zhu, and Qiang Boqing).

References

Bowcock, A.M., Ruiz-Linares, A., Tomfohrde, J., Minch, E., Kidd, J.R., and Cavalli-Sforza, L.L. 1994. High resolution of human evolutionary trees with polymorphic microsatellites. *Nature* 368:455-457.

Brooks, A.S. and Wood, B. 1990 . The Chinese side of the story. *Nature (London)* 344:288-289.

Cann, R.L. 1996. Hybrids, mothers, and clades: Who is right? IN:Akazawa, R. and Szathmary, E.J.E. (eds.) *Prehistoric Mongoloid Dispersals*. Oxford: Oxford University Press, pp. 41-51.

Cavalli-Sforza, L.L. and Edwards, A.W.F. 1967 Phylogenetic analysis: models and estimation procedures. Am. J. Hum. Genet. 19:233-257.

Chu, J.Y., Huang, W., Kuang, S.Q., Wang, J.M., Xu, J.J., Chu, Z.T., Yang, Z.Q. et al. 1998. Genetic relationship of populations in China. *Proc. Natl. Acad. Sci. USA* 95:1763-1768.

Deka, R., Jin, L., Shriver, M.D., Yu, L.M., DeCroo, S., Hundrieser, J., Bunker, C.H., et al. 1995. Population genetics of dinucleotide (dC-dA)n.(dG-dT)n polymorphisms in world populations. *Am. J. Hum. Genet.* 56:461-474.

Deka, R., Jin, L., Shriver, M.D., Yu, L.M., Saha., N. Barrantes, R., Chakraborty, R., and Ferrell, R.E. 1996. Dispersion of human Y chromosome haplotypes based on five microsatellites in global populations. *Genome Res.* 6:1177-1184.

Diamond, J.M. 1988. Express train to Polynesia. *Nature* 336:307-308.

Etler, D.A. 1992. Recent developments in the study of human biology in China: a review. *Hum. Biol.* 64:567-585.

Goldstein, D.B., Ruiz-Linares, A., Cavalli-Sforza, L.L. and Feldman, M.W. 1995a. An evaluation of genetic distances for use with microsatellite loci. *Genetics* 139:463-471.

Goldstein, D.B., Ruiz-Linares, A., Cavalli-Sforza, L.L. and Feldman, M.W. 1995b. Genetic absolute dating based on microsatellites and the origin of modern humans. *Proc. Natl. Acad. Sci. USA.* 92:6723-6727.

Grimes, B.F. 1996. *Ethnologue.* Dallas: Summer Institute of Linguistics, 13th edition.

Jin, L. and Su, B. 2000. Natives or immigrants: modern human origin in East Asia. Nature Reviews Genetics. 1:126-133.

Jorde, L.B., Bamshad, m.J., Watkins, W.S., Zenger, R., Fraley, A.E., Krakowiak, P.A., Carpenter, K.D., et al. 1995. Origins and affinities of modern humans: A comparison of mitochondrial and nuclear genetic data. *Am. J. Hum. Genet.* 57:523-538.

Li, T. and Etler, D.A. 1992. New middle Pleistocene hominid crania from Yunxian in China. *Nature (London)* 357:404-407.

Melton, T. Clifford, S. Martinson, J. Batzer, J. Stoneking, M. 1995 Polynesian genetic affinities with Southeast Asian populations as identified by mtDNA analysis. *Am. J. Hum. Genet.* **57**, 403-414.

Nei, M. 1987 *Molecular Evolutionary Genetics.* NY: Columbia University Press.

Qian, Y.P., Qian, B.Z., Su, b., Yu, J.K., Ke, Y.H., Chu, Z.T., Shi, L. et al. 2000. Multiple origins of Tibetan Y chromosomes. *Hum. Genet.* 106:453-454.

Saitou, N. and Nei, M. 1987. The neighbor-joining method: a new method for reconstructing phylogenetic trees. Mol. Bio. Evol.4(4):406-425.

Shriver MD. Jin L. Ferrell RE. Deka R. 1997 Microsatellite data support an early population expansion in Africa. Genome Research. 7(6):586-591.

Slatkin, M. 1995. A measure of population subdivision based on microsatellite allele frequencies [published erratum appears in Genetics. *Genetics* 139:457-462.

Su, B., Jin, L., Underhill, P., Martinson, J., Saha, N., McGarvey, S.T., Shriver, M.D. et al. 2000. Polynesian origins: Insights from the Y chromosome. *Proc. Natl. Acad. Sci USA* 97:8225-8228.

Su, B., Xiao, J.H., Underhill, P., Deka, R., Zhang, W.L., Akey, J., Huang, W. et al. 1999. Y-chromosome evidence for a northward migration of modern humans into East Asia during the Last Ice Age. *Am. J. Hum. Genet.* 65:1718-1724.

Wang, L. 1986. Secular change and geographical variation in Chinese Neolithic and modern inhabitants: A statistical study of craniometric traits. *Acta Anthropol. Sin.* 5:243-258.

Weng, Z., Yuan, Y. and Du, T. 1989. Analysis of the genetic structure of human populations in China. *Acta Anthropol. Sin.* 8:261-168.

Zhang, H. 1988. Distribution of dermatoglyphic parameters in fifty-two Chinese populations. *Acta Anthropol. Sin.* 7:39-45.

Zhang, Z.B. 1988. An analysis of the physical characteristics of modern Chinese. *Acta Anthropol. Sin.* 7:314-323.

Zhao, T.M., Zhang, G., Zhu, Y., Zheng, S., Liu, D., Chen, Q. and Zhang, X. 1986. The distribution of immunoglobulin Gm allotypes in forty Chinese populations. *Acta Anthropol. Sin.* 6:1-8.

Zhao, T.M. and Lee, T.D. 1989.Gm and Km allotypes in 74 Chinese populations: A hypothesis of the origin of the Chinese nation. *Hum. Genet* 93:101-110.

ORIGINS AND PREHISTORIC MIGRATIONS OF MODERN HUMANS IN EAST ASIA

BING SU
Kunming Institute of Zoology, Chinese Academy of Sciences
Human Genetics Center, University of Texas
School of Life Sciences, Fudan University

LI JIN
Human Genetics Center, University of Texas
School of Life Sciences, Fudan University

Introduction

There are two competing hypotheses on the origin of modern humans: the Out-of-Africa hypothesis and the multiregional hypothesis. Both agree that *Homo erectus* originated in Africa and expanded to Eurasia about one million years ago, but they differ in explaining the origins of modern humans (see reviews in Wilson et al. 1992 and Thorne et al. 1992). One theory claims another Out-of Africa event happened about 100,000 years ago, when anatomically modern humans (*Homo sapiens sapiens*) of African origin would have conquered the world by completely replacing archaic human populations (*Homo sapiens*). The other theory states that independent multiple origins or shared multiregional evolution occurred in the million years since *Homo erectus* came out of Africa (the trellis theory). A compromised version of the Out-of-Africa hypothesis is open to the possibility of minor local contributions (Figure 1).

In the past decade, the Out-of-Africa hypothesis has been supported by extensive genetic evidence as well as archaeological findings (Cann et al. 1987; Stringer and Andrew 1988; Vigilant et al. 1991; Bowcock et al. 1994; Hammer 1995; Tishkoff et al. 1996; Chu et al. 1998; Su et al. 1999; Quintana-Murci et al. 1999; Underhill et al. 2000). However, it was claimed, mostly by archaeologists, that the abundant hominid fossils found in China and other regions in East Asia demonstrate continuity not only in morphological characters but also in spatial and temporal distributions (Brooks and Wood 1990; Li and Etler 1992; Wu and Poirier 1995; Etler 1996). This observation implies the involvement of East Asia in the evolution from *Homo erectus* to *Homo sapiens* and then to *Homo sapiens*

sapiens, and has been used repeatedly to challenge the validity of the Out-of-Africa hypothesis (Wu and Poirier 1995; Etler 1996). Unfortunately, genetic evidence supporting the Out-of-Africa hypothesis is not without dispute and the interpretation of the genetic evidence, coming mostly from mitochondrial DNA studies, has been controversial (Wilson and Cann 1992; Thorne and Wolpoff 1992; Templeton 1992; Relethford and Harpending 1995; Cavalli-Sforza et al. 1996; Ayala 1996; Stoneking and Soodyall 1996; Templeton 1997). Furthermore, the representation of the Chinese samples (mostly southern Chinese originally from Canton) was questioned by examining the competing hypotheses of modern human origin in East Asia (Vigilant et al. 1991; Bowcock et al. 1994; Hammer et al. 1995; Tishkoff et al. 1996) considering the presence of a large number of diversified ethnic populations in this region. On the other hand, although the expansion of modern humans into Europe, the Americas, and Oceania (Australia and New Guinea) are now relatively well characterized (Cavalli-Sforza et al. 1996), little was known about the earliest migratory routes by which modern humans spread from Africa to Asia. Therefore, a close examination of East Asians is essential to resolve the long-standing controversy over the origin of modern humans.

A Recent African Homeland of East Asians

1. Autosome markers

The first piece of genetic evidence on the East Asian origin came from the study of autosomal microsatellite loci (Chu et al. 1998). A total of 30 autosomal microsatellite markers were typed in 28 East Asian populations. Using an approach based on the phylogenetic relationships of populations, it was shown that all East Asian populations, including a variety of ethnic minorities from China, form a single cluster rooted by African populations. This observation strongly supports a common African origin of East Asian populations (Piazza 1998). In addition, the genetic divergence between northern and southern East Asians was confirmed since they belong to distinct clusters in the phylogenetic tree. The phylogenetic analysis suggested an entry of modern humans from the south followed by a

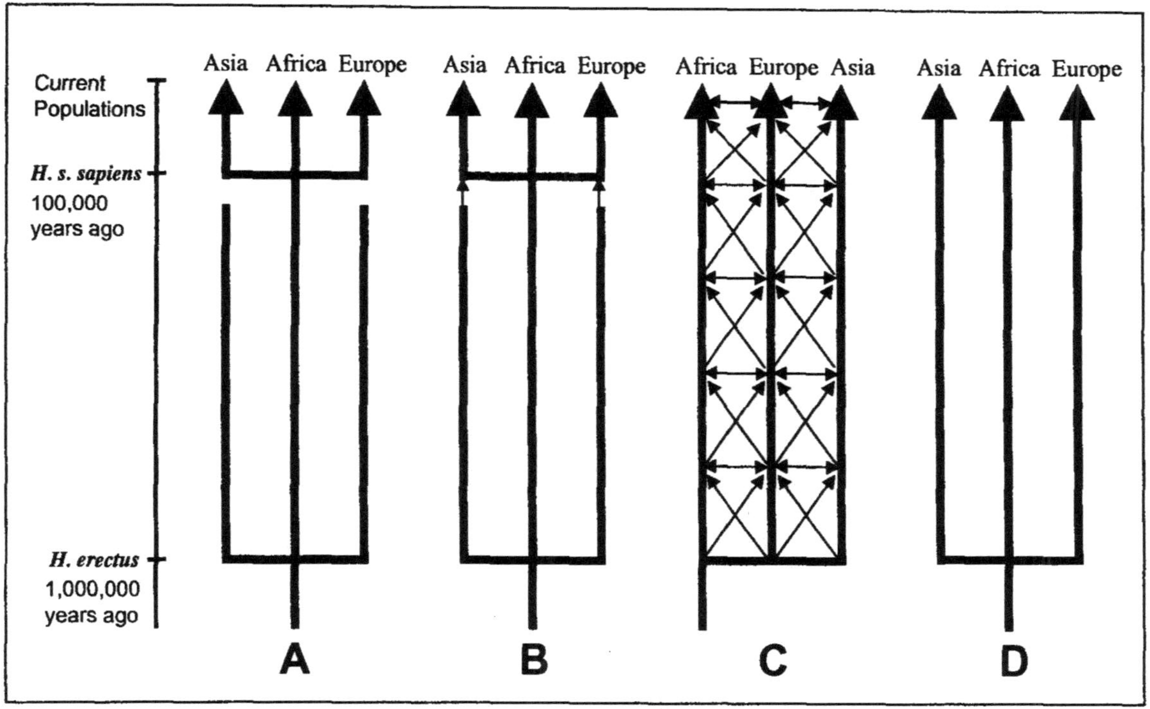

Figure 1. Four different hypotheses on modern human origins. (A) The Out-of-Africa hypothesis with complete replacement. (B) The Out-of-Africa hypothesis with trivial local contributions. (C) The multiregional hypothesis assuming a network of gene exchanges between continental populations (known as the *trellis* model). (D) The multiregional hypothesis of independent origins of modern humans.

northward migration leading to the peopling of East Asia. However, the phylogeny presented in this study has relatively weak statistical support and falls short in providing an unequivocal picture of the prehistoric migrations of modern humans in that region.

A recent study of the biallelic markers on chromosome 21 also supports the African origin of East Asians (Jin et al. 1999). A 565bp DNA fragment near the MX1 gene was found to contain nine polymorphic sites in human populations. No recombination and recurrent mutations (e.g. A to G and then back to A) were observed in this region. Five East Asian populations were included in this study, suggesting a migration from Africa to East Asia.

2. *Y chromosome haplotypes*

A clear picture of modern humans origin and migrations in East Asia has been drawn with Y chromosome markers. Although researchers recognized the utility of Y chromosome markers in reconstructing human history (Jobling and Tyler-Smith 1995), the use of Y chromosome markers had been hampered until the introduction of denaturing high-performance liquid chromatography, an efficient mutation-detection technique (Underhill et al. 1997; Oefner and Underhill 1995; Oefner and Underhill 1998). A large number of Y chromosome biallelic markers have been identified in the past few years (Underhill et al. 1997; Thomson et al. 2000; Shen et al. 2000). We used a set of 19 highly informative Y chromosome biallelic markers, developed at Stanford University, to reveal the genetic structure of the paternal lineages in East Asia. A large collection of population samples covers 21 Chinese ethnic-minorities, 22 provincial Han Chinese, 3 northeast-Asian populations and 5 southeast-Asian populations as well as 12 non-Asian populations from Africa, America, Europe and Oceania (Su et al. 1999). The well represented samples and the highly informative Y chromosome markers led to the conclusion that southern East Asian populations are much more diverse than their northern counterparts. The extremely diverse haplotypes in Thai and Cambodian populations indicates that mainland Southeast Asia might be the first settlement of modern humans of African origin. Furthermore, this initial settlement was estimated to have occurred 18,000—60,000 years ago. A northward diaspora, coinciding with the retreating glacier of the Last Ice Age, resulted in the peopling of China and extended far north into Siberia. The genetic difference observed between northern and southern East Asian populations is probably attributable to the geographic separation after a bottleneck event during the northward migration (Su et al. 1999).

A total of 17 haplotypes derived from 19 Y chromosomal biallelic markers are present in more than 1,000 East Asian individuals with diverse ethnic background (Table 1). No recurrent mutations were observed and the phylogenetic relationships among the haplotypes are unequivocal due to the non-recombinant nature of Y chromosomes (Figure 2). Therefore, the geographic distribution of those haplotypes simply reflects the demographic history of the populations assuming no selection. H1 is the ancestor haplotype and H2-H17 are the derived ones. The majority (~70%) of East Asian Y chromosomes (H6-H12) are East Asian specific but absent in other world populations, indicating a common origin of East Asians. Those haplotypes specific to East Asians share a C to G mutation at locus M9 that is also prevalent in other world populations except in Africa (Underhill et al. 1997). H5 and H13-H17 also belong to this lineage. H4 was derived from H1 but is ancestral to the M9G haplotypes (H5-H17) and is defined by a C to T mutation at locus M89. This mutation has a non-trivial contribution (~4% on average) to East Asians, especially the northerners (e.g. 20% in Hui, 11% in Manchurians and 8.3% in Mongolians). The H2 and H3 haplotypes are also derived from H1 and contain the insertion of a 300 bp Alu sequence (Y chromosome Alu polymorphism, YAP; M1 in Figure 1) (Hammer et al. 1995), which are predominant in Tibetan and Japanese populations but generally absent in other East Asians. Therefore, the majority of East Asian Y chromosomes (H2-H17) can be traced back to an African ancestor. However, since H1 is the ancestral haplotype and the H1s (about 10%) in East Asians are indistinguishable from those in Africans by the 19 biallelic markers, they became the only possibility that may contain Y chromosomal lineages representing an ancient local origin.

Recently, up to 98 new Y chromosomal variants were reported in the coding regions of four Y chromosome genes (Thomson et al. 2000; Shen et al. 2000). The phylogenetic tree constructed with 56 of the 98 mutations in world populations is consistent with the Out-of-Africa hypothesis (Cann et al. 1987; Vigilant et al. 1991). It reveals an extreme geographic structure, with the oldest clades representing Africans and the younger ones

Table 1. Y chromosome haplotype frequency distributions in linguistically and culturally diversified East Asian populations

Region	Population	Language`	H1A	H1B	H2	H3	H4	H5	H6	H7	H8	H9	H10	H11	H12	H13	H14	H15	H16	H17	Size
	Locus		M130	M48	M1	M15	M89	M9	M122	M7	M134	M119	M103	M95	M88	M120	M45	M3	M17	M5	
North E. Asia	Buryat	a	75				25														4
	Ewenki	a		50				13	13		25										8
	Manchurian	a	17				11	22	28		17	5.6									18
	Mongolian	a	13	46	4.2		8.3	13	4.2		4.2	4.2							4.2		24
	Hui	a	10		5		20	30			20						10		5		20
	Korean	b						57			43										7
	Japanese	b	21		28			21	17		10	3.4									29
East Asia	Shandong Han	c	9.4				3.1	18.8	28.1		28.1	9.4				3.1					32
	Henan Han	c	7.1				3.6	25.0	32.1		14.3	10.7				7.1					28
	Northern Han	c	9.1					22.7	27.3		27.3	9.1				4.5					22
	Anhui Han	c	13.6					18.2	27.3		18.2	18.2				4.6					22
	Zhejiang Han	c	12.0					6.0	24.0		26.0	26.0		6.0							50
	Jiangsu Han	c	7.3		1.8		1.8	18.2	23.6	3.6	21.8	16.4		3.6		1.8					55
	Shanghai Han	c	6.7		3.3			16.7	23.3	3.3	16.7	26.7				3.3					30
	Hubei Han	c	5.6					11.1	27.8	5.6	33.3	16.7									18
	Sichuan Han	c			7.1			28.6	7.1	7.1	35.7	7.1		7.1							14
	Jiangxi Han	c	4.8		4.8		9.5	19.1	19.1		23.8	14.3		4.8							21
	Hunan Han	c						13.3	33.3		26.7	13.3		13.3							15
	Fujian Han	c	7.7					7.7	38.5		38.5				7.7						13
	Yunnan Han	c	11.1				3.7	3.7	18.5		55.6	3.7		3.7							27
	Guangxi Han	c	33.3									66.7									3
	Guangdong Han	c							40.0		26.7	20.0			6.7	6.7					15
	Tibetan-Zang	d	8.7		24	17	4.3		4.3		35			2.2		2.2	2.2				46
	Tibetan-Khamba	d			15	30	3.7	15	7.4		30										27
	Jingpo	e									100										5
South Asia	Assam	e						15.0			85.0										20
	Rabha	e	11.8				5.9	5.9			76.5										17
	Naga	e							33.3		66.7										3
	Adi	e									100										5
	Nishi	e						11.1			88.9										9
	Apatani	e					20.0				80.0										5
South E. Asia	Tujia	f	10					20	30	10		20			10						10
	Yi	f				14		43	21		7.1			14							14

Table 1 (continued)

Region	Population Locus	Language*	H1A M130	H1B M48	H2 M1	H3 M15	H4 M89	H5 M9	H6 M122	H7 M7	H8 M134	H9 M119	H10 M103	H11 M95	H12 M88	H13 M120	H14 M45	H15 M3	H16 M17	H17 M5	Size
South E. Asia	Jino	f	17				5.6	39	11		22			5.6							18
	Lahu-Yunnan	f	8.1				19	38	16		11			8.1							37
	Lahu-Thailand	f	7.7					23	15		23			7.7	7.7	7.7			7.7		13
	Bai	f	15			7.7		15	31		31										13
	Naxi	f						78						17			5.5				18
	Akha	f													31	69					13
	Lisu-Thailand	f							40		30			10	20						10
	Karen	g						37	11		53										19
	She	h	18					9.1	18	27	18			9.1							11
	Yao-Nandan	h	50					20		30											10
	Hmong	h				33	33		33												3
	Yao-Thailand	h	7.1				7.1	21	14		36				14						14
	Cambodian	i	3.8			3.8	12	12	3.8		15	3.8	3.8	23	12		3.8		3.8		26
	Wa	i	8.5				2.1	21	32		34			2.1							47
	Blang	i	20					20						60							5
	Deang	i								50	50										2
	Sakai	i					20		60					20							5
	Muong	i					8.3	8.3	25		17			42							12
	Kinh	i					6.7	6.7	33		13	6.7		33							15
	Yao-Jinxiu	j				50			10					40							10
	Zhuang	j	3.6			3.6	7.1	3.6	3.6		25	18		25	11						29
	Dong	j	20						10		20	20	10	20							10
	Li	j									9.1	27		55	9.1						11
	Dai	j	5				7.5	23	13		30			5	18						40
	N. Thailand	j			20				5		30			20	20		5				20
	NE Thailand	j					5	5	5	5		5	5	45	20		5				20
	Ahom	j					20				50		10				20				10
	Bunun	k										11	67		22						9
	Atayal	k							29	4.2	4.2	54	8.3								24
	Yami	k										25		75							8
	Paiwan	k									18	55	27								11
	Ami	k										100									6
Oceania	Oceanian	l	16.0				2	40.0									4			38.0	100

*a=Altaic, b=Isolate, c=Sino-Tibetan, Chinese, Mandarin, d=Sino-Tibetan, Tibeto-Burman, Bodic, e=Sino-Tibetan, Tibeto-Burman, Baric f=Sino Tibetan, Tibeto-Burman, Burmese-Lolo, g=Sino-Tibetan, Tibeto-Burman, Karen, h=Hmong-Mien, i=Austro-Asiatic, j=Daic, k=Austronesian, l=Non-Austronesian

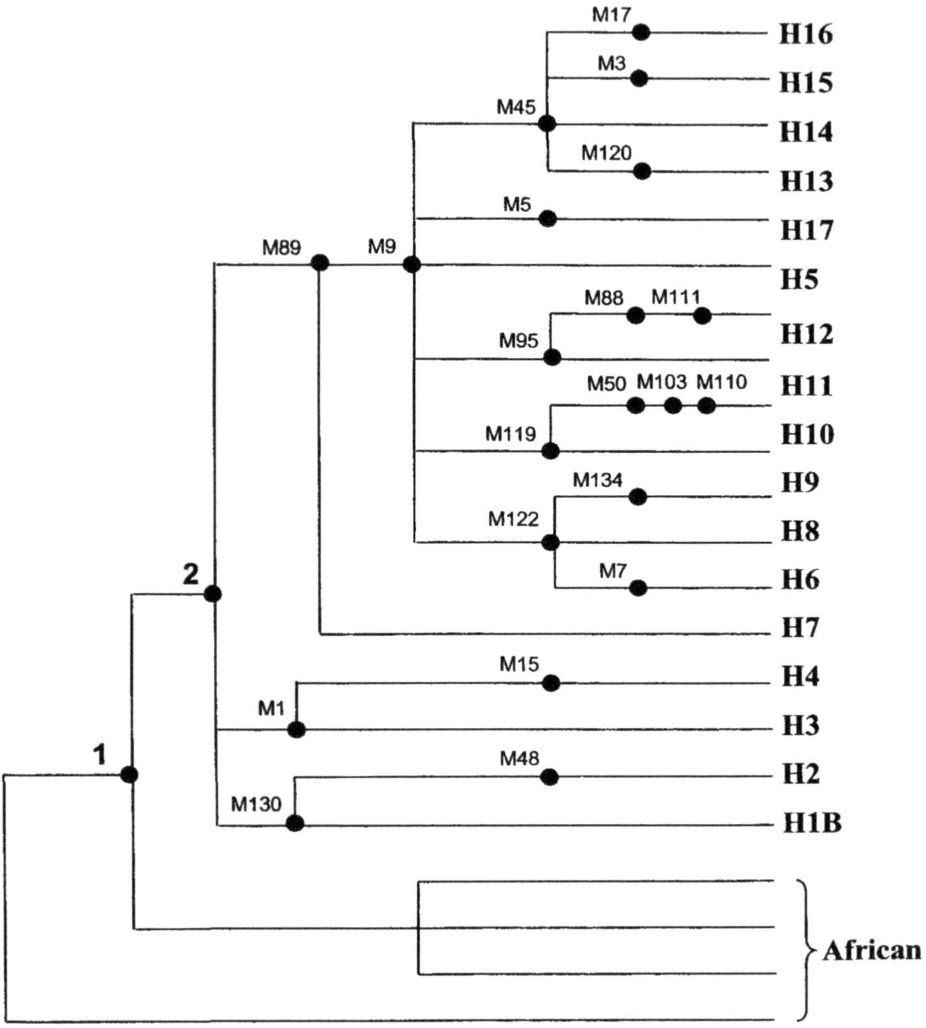

Figure 2. The phylogenetic tree of Y chromosome haplotypes observed in East Asian populations (H1-H17) (Su et al. 1999). The genetic loci associated with the haplotypes are labeled beside the branches. Mutations 1 and 2 are two ancient polymorphisms reported by Shen et al. (2000).

representing some Africans and all non-Africans. In other words, all the non-African Y chromosomes are derived from a small number of Africans. The estimation of the common ancestor of the Y chromosomes is less than 80,000 years (Thomson et al. 2000). Two critical mutations, defined as mutation 1 and 2, are believed to be associated with the Out-of-Africa migratory event and occurred quite recently (40,000-47,000 years ago) (Thomson et al. 2000). When the East Asian data is incorporated into the global tree, all the H1 haplotypes found in East Asia fall into the younger clade defined by mutation 2, a derived polymorphism from mutation 1 which is only polymorphic in Africa (Figure 2). Therefore, all Y chromosomes sampled from East Asia (1,135 Y chromosomes) are directly derived from African lineages. Hence, the replacement of modern humans of African origin in East Asia is complete. A further study with a large sample size of more than ten thousand individuals in 165 populations gave a strong confidence to the completeness. Three critical loci, M1, M89 and M130 defining three main male lineages under mutation 2, were typed in 12,169 samples representing 165 ethnic populations all across Asia (Table 2). As expected, all the individuals typed have one of the derived alleles at these three loci. In other words, no ancient local haplotypes were found in Asian populations leaving the multiregional hypothesis a remote possibility.

Controversy in Interpreting Genetic and Archaeological Evidence

The Out-of-Africa hypothesis has been subjected to two major criticisms from the school of multiregionalists. The first is the claim of misinterpretation of the multiregional hypothesis as independent origins. A network (trellis) model is evoked allowing frequent gene exchanges between continental populations since *Homo erectus* came out of Africa about one million years ago, causing shared multiregional evolution across the human range (Wolpoff et al. 2000). It was claimed that the extensive genetic data, mainly the mitochondrial DNA data supporting the Out-of-Africa hypothesis could also be explained under this version of multiregional hypothesis. However, the *trellis model* fails to explain the haplotype data from Y chromosomes. As discussed above, evidence from Y chromosome biallelic markers unambiguously support the Out-of-Africa model, since all of the East Asian specific haplotypes are derived from a small number of African

Table 2a. The distribution of three Y chromosome polymorphisms in a sample of 10,000 male individuals in China

Regions	N_Pop	N_Ind	M130T	YAP+	M89-C
Central Asia	13	173	14.45	2.31	83.24
Central Siberia	5	107	33.64	0.93	65.42
Okhotsk/Amur	6	123	64.00	0.00	36.00
Kamchatka/Chukotka	4	102	28.40	0.00	71.60
Northern East Asia	17	578	7.27	6.74	85.99
Northern Han Chinese	14	4592	4.16	2.29	93.55
Southern Han Chinese	13	5127	1.91	0.80	97.29
Taiwanese	5	58	0.00	0.00	100.00
Southeast Asia	37	620	6.61	3.55	89.84
Indonesia/Malasia	25	355	6.20	0.00	93.80
Polynesia/Micronesia	11	113	20.36	0.89	78.76
Papuan/New Guinea/Melanesia	7	120	14.17	0.00	85.83
Northeast India	6	59	3.39	0.00	96.61
Total	163	12,127			

Table 2b. The 163 populations typed in this study

Okhotsk/Amur	Tibetan-Yunnan	Blang-1	Makassar
Downriver Negidal	Tu	Blang-2	Malay CB
Nivkh	Uyghur	Cambodian	Malay KM
Okhotsk Evenk	**Northern Han Chinese**	Dai-1	Malay(Pakanbaru)
Udegey	Beijing	Dai-2	Manado
Ulchi/Nanai	Gansu	Deang	Minangkabau
Upriver Negidal	Hebei	Dong	Nias
Central Siberia	Heilongjiang	Hmong	Orang Asli-1
Buryat-1	Henan	Jino	Orang Asli-2
Buryat-2	Hubei	Karen	Palembang
Tofalar	InnerMongolia	Kinh	Sakai
Tuvan	Jilin	Lahu-1	Sasak
Yenisey Evenk	Liaoning	Lahu-2	Sumba
Kamchatka/Chukotka	Qinghai	Lahu-3	Sumbawa
Chukchi	Shaanxi	Lahu-4	Tengger
Itelman	Shandong	Li	Torajan
Koryak	Shanxi	Lisu	**Poly/Micronesia**
Siberian Eskimo	Tianjin	Muong	American Somoan
Central Asia	Xinjiang	N. Thailand	Guam
Bukharan Arab	**Southern Han Chinese**	Naxi	Kapingamarangi
Crimean Tatar	Anhui	NE Thailand	Kribati
Dungan (Hui)	Fujian	She	Majuro
Eastern Uzbec	Guangdong	So	Nauru
Iranian	Guangxi	Tujia	Palau
Karakalpak	Guizhou	Wa-1	Pohnpei
Kazak	Hubei	Wa-2	Tonga
Khorezmian Uzbek	Hunan	Wa-3	Truk
Kyrgyz	Jiangsu	Yao	West Somoan
Sinte Romani (Gypsy)	Jiangxi	Yao-Jinxiu	**Papuan&NG**
Tajik	Shanghai	Yao-Nandan	Australian Aborigine
Turkmen	Sichuan	Yi	Bankes&Torres
Uighur	Yunnan	Zhuang	Maewo
Northern East Asia	Zhejiang	**Indonesia/Malasia**	Nasioi Melanesian
Ewenki	**Taiwan Aborigines**	Alor	New Guinean-1
Hui	Ami	Bali	New Guinean-2
Japanese	Atayal	Bangka	Santo
Jingpo	Bunun	Batak-1	**NE India**
Kazak-Xingjiang	Paiwan	Batak-2	Adi
Korean	Yami	Bugis	Apatani
Manchurian	**Southeast Asia**	Dayak	Assam
Mongolian	Ahom	Irian	Naga
Mongolian	Aini	Java-1	Nishi
Sala	Akha	Java-2	Rabha(Assam)
Tibetan-Qinghai	Bai-1	Kaili	
Tibetan-Tibet	Bai-2	Kota Kinabalu	

ancestors and show no ancient local contributions. In particular, all the mutations found outside Africa are younger than 50,000 years and were derived from Africa (Thomson et al. 2000). On the other hand, if extensive gene flow occurred between continental populations during the past one million years, the ancient Y chromosome haplotypes seen in African populations, or even much older haplotypes, would also be expected in East Asia, which apparently is not the case after an extensive search in more than 10,000 samples from 165 ethnic populations. The only remaining genetic evidence that seems to be irreconcilable with the Out-of-Africa hypothesis comes from a few large fragment sequencing studies on autosomal and X chromosomal genes (Harding et al. 1997; Kaessmann et al. 1999; Harris and Hey 1999). The age estimations for a common ancestor of these genes range from 535,000 to 1,860,000 years, which is much older than those for mtDNA and Y chromosomes and were considered to be in favor of a multiregional origin. A recent study on a 10kb noncoding region of chromosome 22 has a similar estimation (~1.29 million years) (Zhao et al. 2000). However, this difference in age estimation only reflects the difference in the effective population sizes between Y chromosome/mtDNA and X chromosome/autosome (3-4 times as many as the former) in the presence of bottleneck events associated with the outbound migrations from Africa (Fay and Wu 1999). In other words, the age estimations based on X chromosomal and autosomal regions are compatible with both competing hypotheses; therefore, they are not useful statistics in distinguishing the two competing hypotheses.

The second argument comes from archaeological findings in East Asia, especially in China. As mentioned above, a continuous evolutionary chain has been proposed to connect *Homo erectus* and *Homo sapiens* in China and this was used as a promising evidence to support an independent origin of modern humans in East Asia or the *trellis model* (Wu and Poirier 1995; Etler 1996). However, in a close examination of the collection of hominid fossils in China, we found a nontrivial gap between archaic humans (*Homo sapiens*) and modern humans (*Homo sapiens sapiens*) in terms of temporal continuity (Figure 3). All the *H. sapiens* fossils are at least 100,000 years old while all the *H. s. sapiens* fossils are younger than 40,000 years (with most falling between 10,000 and 30,000 years). In other words, no hominid fossils that can be dated from 100,000 to 40,000 years ago have yet been found in East Asia. This finding is particularly anomalous given the abundance of either earlier or later fossil records that have been found in this area (Wu and

Poirier 1995; Etler 1996). The extensive duration of the temporal discontinuity of the fossil records in China and the distinctive morphological characters of the hominid fossils found before and after would strongly argue against any casual explanation that this gap is attributable to a "missing link." Instead, the extinction of local archaic humans and later peopling of modern humans from Africa seems to be a more reasonable explanation. Interestingly, this fossil gap coincides with the Last Ice Age, during which modern humans of African origin arrived at East Asia in the south.

The Northward Diaspora in East Asia

The validation of the Out-of-Africa hypothesis requires an understanding of the migratory routes of modern humans from Africa to East Asia. One observation related to the prehistoric population movements is a substantial distinction between northern and southern East Asian populations observed in the analyses of both genetic markers and physical characteristics (Du and Xiao 1997; Zhao et al. 1986; Zhao and Lee 1989; Weng et al. 1989; Zhang 1988; Zhang 1988). Three models have been proposed to interpret the observation. The first model postulates a north-to-south migratory pattern, which led to admixture with Australoids in the south (Du and Xiao 1997). The second model suggests a southern origin and northward migration of East Asians (Chu et al. 1998; Su et al. 1999). The third one assumes that the ancestors of the northern and southern populations arrived in East Asia separately. Genetic data collected on Y chromosomes and autosomal microsatellite markers support the second model. According to the distribution of Y chromosome haplotypes in East Asian populations, southern populations are much more diverse than northern populations. The majority of Y haplotypes found in northern populations were derived from a subset of the southern ones, which make the third model untenable (though the presence of YAP+ and H4 may indicate a contribution from a migration originating in Central Asia that arrived in East Asia much later). Furthermore, a principal component analysis showed that all northern populations cluster together at the upper-right corner and are well separated from the southern East Asian populations, which are far more diversified than the northern populations (Figure 4). The proposed possible admixture between East Asians and Australoids (first model) is less likely since no East Asian specific haplotypes are present in the non-Austronesian speaking

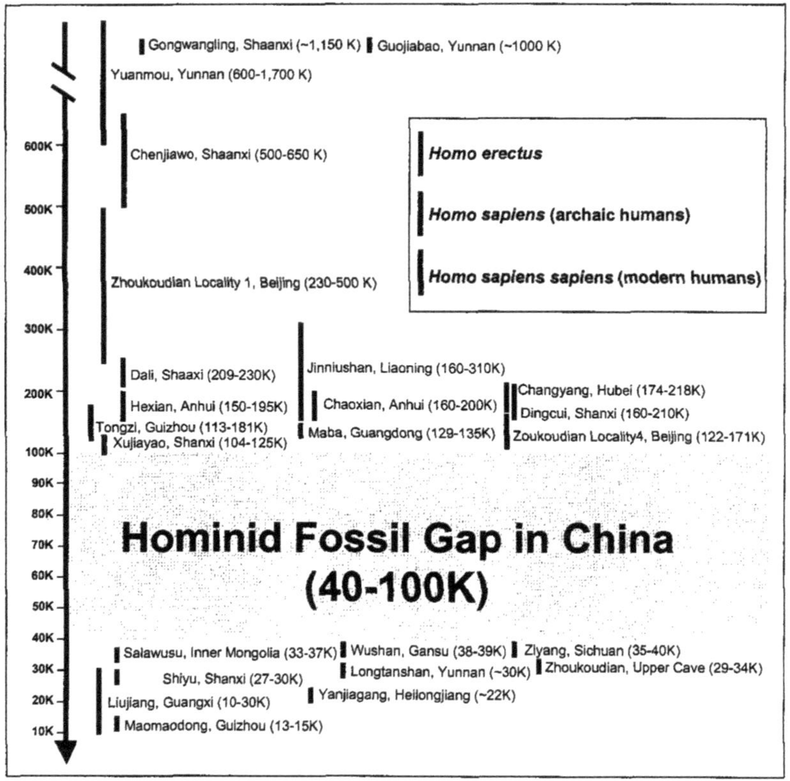

Figure 3. The hominid fossil records found in China so far. The age estimations and the geographic locations of those fossils are labeled beside the time bars. A non-trivial gap exists between 100,000 and 40,000 years ago without any hominid fossils found so far.

populations sampled from New Guinea Highlanders and Melanesia Nasioi (presumably Australoids). In addition, the non-Austronesian specific haplotype (H17) was not observed in East Asians (Su et al. 1999, 2000). Therefore, the major migratory pattern of modern humans in mainland East Asia is from south to north. During this northward diaspora, a group of people reached the Upper Yellow River basin about 30,000 years ago, where they created one of the earliest Chinese cultures and eventually migrated to the Himalayan plateau about 6,000 years ago (See Chapter 5, for the migration study of Sino-Tibetan populations). In addition, after modern humans from Africa settled in mainland Southeast Asia, other than the northward expansion to China and Siberia, they also migrated southward to explore the new habitats, resulting in the peopling of island Southeast Asia and finally triggering the express train to Polynesia (see Chapter 12 for the study of Polynesian origins).

East Asian Contributions to the New World

Besides the East Asian specific Y chromosome haplotypes showing a northward migration, the other haplotypes are also storytellers. Our further genetic dissection with a new Y chromosome marker (Bergen et al. 1999) showed that all the H1 haplotypes in East Asians share a C to T mutation at locus RPS4Y (M130 in Figure 2). The M130T allele is strictly distributed in Asian and Oceanian populations but absent in Africa and Europe. It has a moderate occurrence in North America (e.g. in Na-Dene speaking populations) but is absent in Central and South America. The geographic distribution of M130T suggests that this mutation is Asian and Oceanian specific and probably occurred somewhere in Asia after the "Out-of-Africa" event (Bergen et al. 1999).

The frequency of M130T varies drastically among different Asian populations (Table 1). In Southeast Asian populations, it ranges from 0 to 25%. It is absent in the Yi, Li, North Thai, Northeast Thai, Malay, Orang Asli and aboriginal Taiwanese, but with moderate frequencies in southern ethnic groups in China, including the Tujia, Yao, Dong and She. In the Han Chinese, the M130T frequency is low: 4.1% in northern Han and 1.9% in southern Han (Table 2). Interestingly, in the northern part of East Asia, the M130T shows a moderate frequency in the Manchurian, Japanese and Korean, and becomes the predominant allele in most of the Altaic speaking

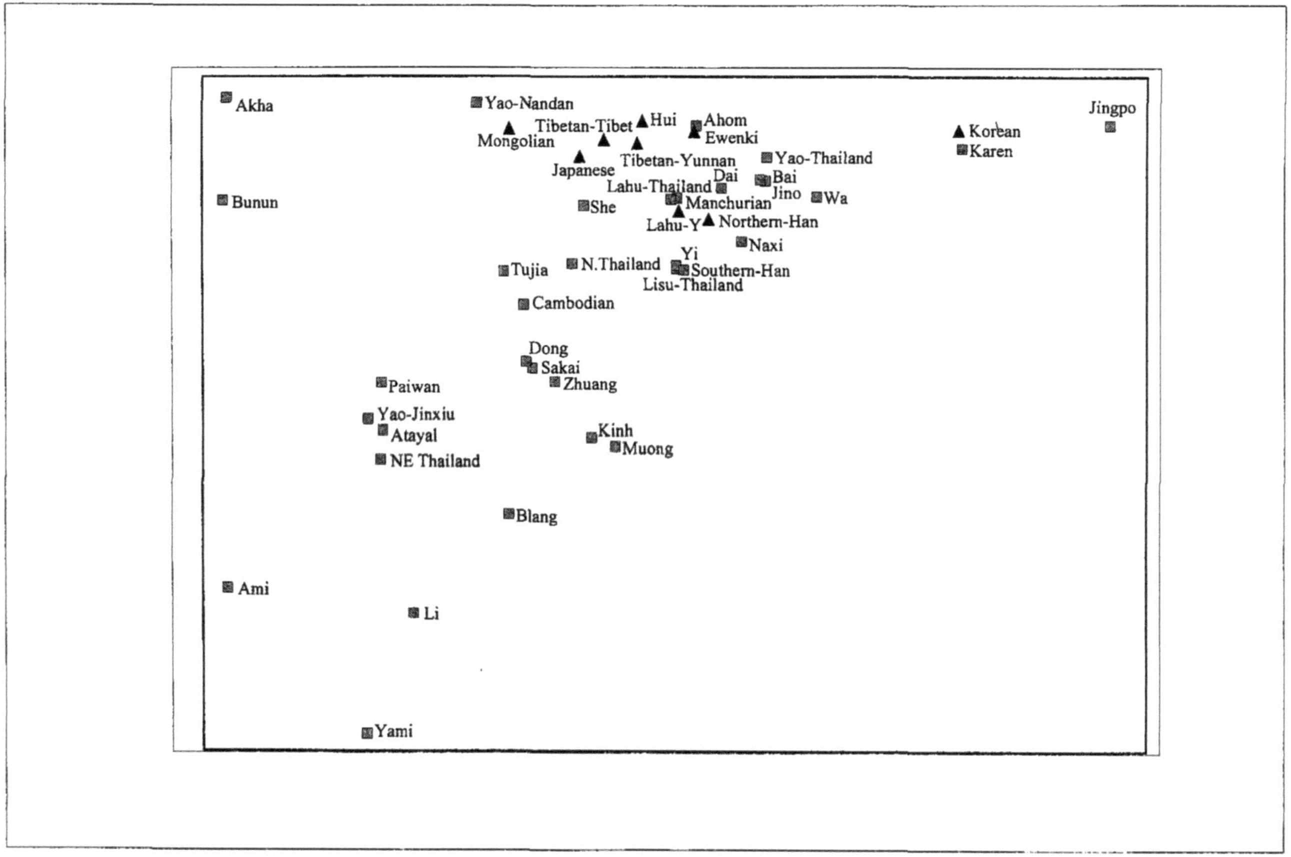

Figure 4. Principal component analysis of 43 East Asian populations. The triangles refer to northern populations while the squares are southern populations. This map accounts for 45% of the original genetic variation.

populations, including Mongolian and most of the Siberian populations except those in the far northeastern region (Table 2).

The M48G, another polymorphism derived from the M130T allele, is only prevalent in Mongolian and Siberian populations without showing any trace in other East Asian populations (Table 3). The prevalence of M48G in Siberia and its absence in other parts of East Asia suggests that this mutation probably originated in East Siberia.

The results of the four Y chromosome microsatellite loci (DYS388, DYS390, DYS391, DYS394) provide more detailed information on the relationship of the Asian populations. A total of 105 haplotypes were observed by combining the M48 polymorphism and the microsatellite variations (Table 3). There are only three shared haplotypes between Siberian and East Asian populations, an indication of a certain genetic divergence between these two geographically adjacent populations.

Among the four alleles observed at DYS391, the DYS391-10 is the most common allele in the East Asian M130Ts, and it is also the predominant allele in other world populations. Therefore, DYS391-10 is likely the ancestral allele at DYS391 (de Knijff et al. 1997). Surprisingly, among all the M130T individuals analyzed, the DYS391-9 allele (the allele containing 9 repeat units) is highly prevalent in Mongolians (78.6%, 12/14), Siberians (90.3%, 65/72) and northern Amerindians (93.3%, 14/15), but generally absent in East Asians, Southeast Asians, Pacific Islanders and Oceanians (Table 3). This contrast in distribution suggests (1) the DYS391-10 allele is the allele ancestral to the DYS391-9 allele in the background of M130T, (2) the M130T in North Asia originated in East Asia, (3) a founder event occurred during this migration which is clearly marked by the haplotype M130T carrying the DYS391-9 allele, and (4) the M130T in North America came from those in Siberia of East Asian origin. The exclusive presence of M48G in Siberia and Mongolia further confirms the aforementioned conclusions. In addition, all the M48G individuals are DYS391-9, indicating the M48 A to G mutation might happen in a DYS391-9 background. There are 5 out of 63 Northern Hans having the DYS391-9 alleles, an indication of probably recent gene flows.

For the individuals carrying the M130T allele, the diversity (in the form of the variance of repeat numbers) of DYS391-9 and DYS391-10 was estimated at the DYS388, DYS390 and DYS394. The variances of DYS391-10 at the three loci are 4 times larger than those of DYS391-9 (6.887 versus 1.626), which mostly is present in Siberian populations and American Indian

Table 3. The Y microsatellite haplotype distribution in 247 M130T individuals in North Asia, East Asia and Oceania

Hap	Size	391	M48	388	390	394	NA	NES	ES	SES	CS	EV	MG	KR	JP	MA	HU	NH	SH	SA	SEA	PY	MC	OC
1	1	9	\	12	24	15	1																	
2	1	9	\	13	24	15	1																	
3	12	9	\	13	24	16	12																	
4	1	9	\	13	24	17								1										
5	1	9	A	11	26	16												1						
6	1	9	A	12	24	14												1						
7	1	9	A	12	25	15			1															
8	1	9	A	12	25	17			1															
9	1	9	A	13	23	15				1														
10	1	9	A	13	23	16				1														
11	1	9	A	13	23	17			1															
12	2	9	A	13	23	18			2															
13	1	9	A	13	24	13			1															
14	1	9	A	13	24	14			1															
15	4	9	A	13	24	16			2						1			1						
16	2	9	A	13	24	17			2															
17	1	9	A	13	25	16							1											
18	2	9	A	13	25	17			1									1						
19	1	9	A	14	24	16												1						
20	1	9	G	11	24	17					1													
21	1	9	G	12	23	17			1															
22	1	9	G	12	24	14			1															
23	12	9	G	12	24	15			12															
24	25	9	G	12	24	16			1	2	22													
25	1	9	G	12	24	17		1																
26	7	9	G	12	25	15			1	6														
27	8	9	G	12	25	17							8											
28	1	9	G	12	25	18							1											
29	1	9	G	12	26	16							1											
30	1	9	G	13	24	16						1												
31	2	9	G	13	25	17						1	1											
32	2	9	G	13	25	18						2												
33	1	10	\	12	24	16								1										
34	1	10	\	12	26	16															1			
35	1	10	\	13	24	16															1			
36	1	10	\	13	24	17								1										
37	1	10	\	13	24	18															1			
38	1	10	A	10	22	16															1			
39	1	10	A	10	24	16												1						
40	1	10	A	10	25	15													1					
41	4	10	A	11	23	15					4													
42	1	10	A	11	24	9															1			
43	2	10	A	11	24	14				2														
44	1	10	A	11	24	16													1					
45	1	10	A	11	26	14												1						
46	1	10	A	12	18	16																		1
47	1	10	A	12	21	16																	1	
48	1	10	A	12	22	16	1																	
49	1	10	A	12	22	17													1					
50	1	10	A	12	23	14												1						
51	10	10	A	12	24	16							1		1			6	2					
52	6	10	A	12	24	17												4	2					
53	2	10	A	12	24	18												1			1			
54	1	10	A	12	25	15													1					

Table 3 (Continued)

Hap	Size	391	M48	388	390	394	NA	NES	ES	SES	CS	EV	MG	KR	JP	MA	HU	NH	SH	SA	SEA	PY	MC	OC
55	3	10	A	12	25	16												1	1	1				
56	2	10	A	12	25	17												1	1					
57	1	10	A	13	20	16																		1
58	5	10	A	13	22	16												1			4			
59	1	10	A	13	22	17													1					
60	1	10	A	13	22	18													1					
61	1	10	A	13	24	9															1			
62	1	10	A	13	24	12											1							
63	1	10	A	13	24	14													1					
64	11	10	A	13	24	16										1		8	1					1
65	7	10	A	13	24	17												3	4					
66	3	10	A	13	24	18												2	1					
67	1	10	A	13	24	20												1						
68	1	10	A	13	25	14									1									
69	7	10	A	13	25	16											1	1	2		3			
70	1	10	A	13	25	17												1						
71	1	10	A	13	26	12												1						
72	1	10	A	13	26	16									1									
73	2	10	A	13	26	17												2						
74	1	10	A	13	27	17													1					
75	1	10	A	14	20	16																		1
76	1	10	A	14	22	17																		1
77	1	10	A	14	24	16															1			
78	1	10	A	14	25	16							1											
79	1	10	A	14	26	17												1						
80	1	10	A	15	20	16																		1
81	1	10	A	15	21	10																1		
82	11	10	A	15	21	17																10	1	
83	2	10	A	15	21	18																2		
84	2	10	A	15	22	17																2		
85	1	10	A	15	25	16													1					
86	1	11	A	10	23	16															1			
87	2	11	A	11	24	16												2						
88	1	11	A	12	24	12												1						
89	2	11	A	12	24	16													2					
90	1	11	A	12	24	17												1						
91	1	11	A	12	26	9															1			
92	1	11	A	12	26	17												1						
93	3	11	A	13	20	16																		3
94	2	11	A	13	20	17																		2
95	1	11	A	13	22	16															1			
96	1	11	A	13	22	17																	1	
97	3	11	A	13	24	16												1	2					
98	2	11	A	13	24	17												1	1					
99	2	11	A	13	25	16												1	1					
100	2	11	A	13	25	17												1	1					
101	1	11	A	13	29	17																		1
102	1	11	A	14	24	13												1						
103	3	11	A	15	21	17																	3	
104	1	11	A	15	27	16													1					
105	1	12	A	13	20	17																		1
Total	**247**						**15**	**1**	**28**	**12**	**27**	**4**	**14**	**3**	**4**	**1**	**2**	**52**	**31**	**1**	**18**	**15**	**6**	**13**

Note: NA = Northern Amerindian; NES = Northeastern Siberian; ES = Eastern Siberian; SES = Southeastern Siberian; CS = Central Siberian; MG = Mongolian; EV = Evenki; KR = Korean; JP = Japanese; MA = Manchurian; HU = Hui; NH = Northern Han Chinese; SH = Southern Han Chinese; SA = South Asia; SEA = Southeastern Asian; PY = Polynesian; MC = Micronesian; OC = Oceanian populations. Furthermore, the variance of DYS391-9/M48A are also larger than those of DYS391-9/M48G for all three microsatellite loci. Hence, the difference of genetic diversity between DYS391-10 and DYS391-9 haplotype groups again supports an East Asian origin of those M130Ts in Siberian populations.

The hypothesis of an East Asian homeland of Siberian M130Ts is reinforced by ruling out the possibility of an alternative of Central Asian origin. Both M130T and M48G are present in several Central Asian populations with relatively moderate frequency. These populations are geographically close to Siberia, including the Karakalpak, Kazak, Kyrgyz, Tajik, Dungan, Tuva, Uighur and Uzbek (Wells, personal communication). However, based on the microsatellite loci, most, if not all, of the M130Ts were most likely derived from that found in East Siberia. Furthermore, the microsatellite diversity of M130Ts in Central Asians is lower than that in East Asian, and the microsatellite diversity of M48G (most of them are DYS391-9) in Central Asians is also much lower than that in Siberians (data not shown). Therefore, instead of the ancestors, the M130Ts in Central Asian might imply a relatively recent westward expansion starting from Siberia as documented in the literary records.

In America, the M130T has a restricted distribution in North America, mainly the Na-Dene speaking populations (Bergen et al. 1999; Karafet et al. 1999; Lell et al. 2000). There are 4 microsatellite haplotypes in the 15 American Indian M130Ts, and 12 of them share the same haplotype, which shall be either the same as the common one shared by two Siberian individuals, or only one step away from the most dominant haplotype in Siberian populations when M48 is typed (Table 3). Hence, the American M130Ts were from Siberia, and it can eventually be traced back to East Asia.

Considering the restricted distribution of M130T in North America, the arrival of M130T in Siberia might be a relatively later event as compared to the major migration marked by the American Indian specific haplotype H16, the predominant male lineage all across the Americas (Underhill et al. 1996). The age of M130T-M48G was estimated to be about 11,000-18,000 years ago assuming an effective population size of 500-1000 and a generation time of 20 years (refer to Su et al. 1999 for the method of age estimation). This estimation probably reflects the entry into Siberia from East Asia, and it is consistent with the divergence time of the Na-Dene populations (North America) (6,000-10,000 years ago) based on mtDNA data (Torroni et al.

1992), and the divergence time of mtDNA haplogroup B in American Indians (13,500 -17,700 years old), which is much younger than the other mtDNA haplogroups (haplogroup A: 27,241—35,902 years; haplogroup C: 27,300—45,230 years; haplogroup D: 23,535—47,650 years) (Schurr 2000). Interestingly, archaeological findings are also consistent with the connection between North Asia and East Asia. The dental evidence showed that the so-called Sinodont dentition in northern Asian peoples occurred about 20,000 years ago, and a similar dentition pattern predominates among all the Southeast Asian populations that are ancestral to the Sinodont pattern (Turner et al. 1995). Therefore, the M130T distribution in East Asia, Siberia and North America indicates a northward route leading to the second wave of migration to the New World.

Central Asian Admixture in East Asia

Another major feature of the Y chromosomes in East Asia is reflected in the appearance of YAP+ and another ancient polymorphism (C to T at locus M89). The haplotypes associated with these two polymorphisms are H2/H3 (YAP+) and H4 (M89T) (Su et al. 1999). The prevalence of YAP+ and M89T in Central and Mid-eastern Asian populations (Wells, personal communication) and their general absence in South East Asian populations suggests a genetic influence from northwest to the East Asian populations starting possibly in Central Asia. The distribution of YAP+ in East Asia is somewhat mysterious. It was demonstrated that only Tibetan in the west and Japanese in the east show high frequencies of YAP+, with sporadic residual occurrence in the other East Asian populations (Hammer et al. 1995). It is also absent in Siberian populations (Karafet et al. 1999), thus forming a substantial geographic gap in the YAP+ distribution in East Asia. Interestingly, besides Tibetan (42.5%) (Qian et al. 2000) and Japanese (27.6%), another Chinese ethnic group (Yao-Jinxiu) living in southern China has appreciable presence of YAP+ (50%) (Table 1). This might provide a clue for tracing the dispersion of YAP+ Y chromosomes, although more data is needed to solve the puzzle of YAP+ distribution. Compared to YAP+, the distribution of H4 is more understandable. The highest concentration of H4 was found in Central Asia (Wells, personal communication) and its influence expanded to many populations in the mainland with more on the northern populations and much less on the southern populations, but didn't reach

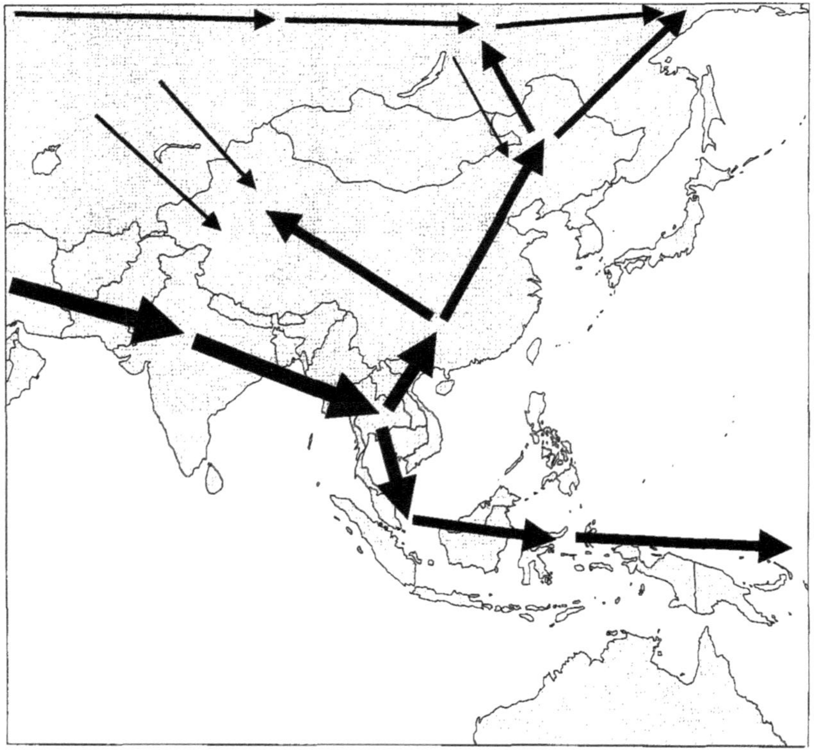

Figure 5. The putative migratory routes of East Asians of African origin in the past 60,000 years as indicated by the dark grey. The first entry of modern humans into the southern part of East Asia occurred about 18,000-60,000 years ago, followed by a northward migration. A southern route also started from mainland Southeast Asia, through Malaysia and Indonesia, eventually marching eastwards to the Pacific Islands. The light grey indicate contributions from Central Asia that arrived in East Asia relatively later.

island populations such as the Taiwanese and Polynesians. Whether YAP+ and M89T from Central Asia came to East Asia in the same migratory wave is subject to further investigations.

Conclusions

In summary, based on the genetic evidence generated so far, particularly from Y chromosome data, modern humans in East Asia have a recent common origin in Africa. The first entry of modern humans into the southern part of East Asia occurred about 18,000-60,000 years ago, followed by a northward migration coinciding with the receding glaciers in that area. A population with a signature of M130T Y chromosome entered Siberia around 11,000-18,000 years ago, probably from northeastern China and crossed the Bering Strait as the second wave of migration to the New World (Figure 5).

References

Ayala, F.J. 1996. Gene lineages and human evolution. *Science* 272:1363-1364.

Bergen, A.W., Wang, C.Y., Tsai, J., Jefferson, K., Dey, C., Smith, K.D., Park S.C. et al. 1999. An Asian-Native American paternal lineage identified by RPS4Y resequencing and microsatellite haplotyping. *Ann. Hum. Genet.* 63:63-80.

Bowcock, A.M., Ruiz-Linares, A., Tomfohrde, J., Minch, E., Kidd, J.R., and Cavalli-Sforza L.L. 1994. High resolution of human evolutionary trees with polymorphic microsatellites. *Nature* 368:455-457.

Brooks, A.S. 1990. Paleoanthropology. The Chinese side of the story. *Nature* 344:288-289.

Cann, R.L., Stoneking, M., and Wilson, A.C. 1987. Mitochondrial DNA and human evolution. *Nature* 325:31-36.

Cavalli-Sforza, L.L. and Piazza, M.P. 1996. *The History and Geography of Human Genes.* Princeton, NJ: Princeton University Press.

Chu, J.Y., Huang, W., Kuang, S.Q., Wang, J.M., Xu, J.J., Chu, Z.T., Yang, Z.Q. et al. 1998. Genetic relationship of populations in China. *Proc. Natl. Acad. Sci. USA* 95:11763-11768.

Du, R., and Xiao, C. J. 1997. Genetic distances between Chinese populations calculated on gene frequencies of 38 loci. *Science in China (Series C)* 40:613-621.

Etler, D.A. 1996. The fossil evidence for human evolution in Asia. *Annu. Rev. Anthropol.* 25:275-301.

Fay, J.C. and Wu, C.I. 1999. A human population bottleneck can account for the discordance between patterns of mitochondrial versus nuclear DNA variation. *Mol. Biol. Evol.* 16:1003-1005.

Hammer, M.F. 1995. A recent common ancestry for human Y chromosomes. *Nature* 378:376-378.

Harding, R.M., Fullerton, S.M. Griffiths, R.C., Bond, J., Cox, M.J., Schneider, J.A. Moulin, D.S. *et al.* 1997. Archaic African and Asian lineages in the genetic ancestry of modern humans. *Am. J. Hum. Genet.* 60:772-789.

Harris, E. and Hey, J. 1999. X chromosome evidence for ancient human histories. *Proc. Natl. Acad. Sci. USA* 96 :3320-3324.

Jin, L., Underhill, P.A., Doctor, V., Davis, R.W., Shen, P.D., Cavalli-Sforza, L.L., and Oefner, P.J. 1999. Distribution of haplotypes from a chromosome 21 region distinguishes multiple prehistoric human migrations. *Proc. Natl. Acad. Sci. USA* 96:3796-3800.

Jobling, M.A. and Tyler-Smith, C. 1995. Fathers and sons: The Y chromosome and human evolution. *Trends in Genetics* 11:449-455.

Kaessmann, H., Heibig, F., Haeseler, A., and Paabo, S. 1999. DNA sequence variation in a non-coding region of low recombination on the human X chromosome. *Nat. Genet.* 22:78-81.

Karafet, T.M., Zegura, S.L., Posukh, O., Osipova, L., Bergen, A., Long, J., Goldman, D. et al. 1999. Y chromosome markers and Trans-Bering Strait dispersals. *Am. J. Phys. Anthropol.* 102:301-314.

de Knijff, P., Kayser, M., Caglia, A., Corach, D., Fretwell, N., Gehrig, C., Graziosi, G., *et al.* 1997. Chromosome Y microsatellites: Population genetic and evolutionary aspects. *Int. J. Legal Med.* 110:134-140.

Lell, J.T., Sukernik, R.I., Starikovskaya, Y.B., Su, B., Jin, L., Underhill, P., and Wallace, D. 2000. The dual origin and Siberian affinities of native American Y chromosomes. *Am. J. Hum. Genet.* (submitted).

Li, T., and Etlter, D.A. 1992. New middle Pleistocene hominid crania from Yunxian in China. *Nature* 357:404-407.

Oefner, P. J. 1998. DNA mutation detection using denaturing high performance liquid chromatography (DHPLC). *Current Protocols in Human Genetics.* NY: Wiley & Sons, Supplement 19,7.10.1-7.10.12.

Oefner, P.J. and Underhill, P.A. 1995. Comparative DNA sequencing by denaturing high-performance liquid chromatography (DHPLC). *Am. J. Hum. Genet.* 57:A266.

Piazza, A. 1998. Towards a genetic history of China. *Nature* 395:636-639.

Qian, Y.P., Qian, B.Z., Su, B., Yu, J.K., Ke, Y.H., Chu, Z.T., Shi, L. et al. 2000. Multiple origins of Tibetan Y chromosomes. *Hum. Genet.* 106:453-454.

Quintana-Murci, L., Semino, O., Bandelt, H.J., Passarino, G., McElreavey, K., and Santachiara-Benerecetti, S. 1999. Genetic evidence of an early exit of *Homo sapiens sapiens* from Africa through eastern Africa. *Nat. Genet.* 23:437-441.

Relethford, J.H. and Harpending H.C. 1995. Ancient differences in population size can mimic a recent African origin of modern humans. *Curr. Anthropol.* 36:667-674.

Schurr, T.G. 2000. Mitochondrial DNA and the peopling of the New World. *Am. Scientists* 88:246-253.

Shen, P., Wang, F., Underhill, P.A., Franco, C., Yang, W.H., Roxas, A., Sung, R. et al. 2000. Population genetic implications from sequence variation in four Y chromosome genes. *Proc. Natl. Acad. Sci. USA* 97:7354-7359.

Stoneking, M. and Soodyall, H. 1996. Human evolution and the mitochondrial genome. *Curr. Opin. Genet. Dev.* 6:731-736.

Stringer, C.B. 1988. Genetic and fossil evidence for the origin of modern humans. *Science* 239:1263-1268.

Su, B., Jin, L., Underhill, P., Martinson, J., Saha, N., McGarvey, S.T., Shriver, M.D. et al. 2000. Polynesian origins: Insights from the Y chromosome. *Proc. Natl. Acad. Sci. USA* 97:8225-8228.

Su, B., Xiao, J.H., Underhill, P., Deka, R., Zhang, W.L, Akey, J., Huang, W. et al. 1999. Y-chromosome evidence for a northward migration of modern humans into East Asia during the Last Ice Age. *Am. J. Hum. Genet.* 65:1718-1724.

Templeton, A.R. 1992. Human origins and analysis of mitochondrial DNA sequences. *Science* 255:7-37.

Templeton, A.R. 1997. Out of Africa? What do genes tell us? *Curr. Opin. Genet. Dev.* 7:841-847.

Thomson, R., Pritchard, J.K., Shen, P.D., Oefner, P.J. and Feldman, M.W. 2000. Recent common ancestry of human Y chromosomes: Evidence from DNA sequence data. *Proc. Natl. Acad. Sci. USA* 97:7360-7365.

Thorne, A. G. 1992. The multiregional evolution of humans. *Sci. Am.* 266:76-79, 82-83.

Tishkoff, S.A., Dietzsch, E., Speed, W., Pakstis, A.J., Kidd, J.R., Cheung, K., Bonne-Tamir, B. et al. 1996. Global patterns of linkage disequilibrium at the CD4 locus and modern human origins. *Science* 271:1380-1387.

Torroni, A., Schurr, T.G., Yang, C.C., Szathmary, E.J. Williams, R.C., schanfield, M.S., Troup, G.A. et al. 1992. Native American mitochondrial DNA analysis indicates that the Amerind and the Nadene populations were founded by two independent migrations. *Genetics* 130:153-62.

Turner, C.G.II. 1993. Shifting continuity: Modern human origin. IN: Brenner, S. and Hanihara, K. (eds.), *The Origin and Past of Modern Humans as Viewed from DNA.*. Singapore: World Scientific, pp 216-243.

Underhill, P.A., Jin, L., Zemans, R., Oefner, P.J., and Cavalli-Sforza, L.L. 1996. A pre-Columbian Y chromosome-specific transition and its implications for human evolutionary history. *Proc. Natl. Acad. Sci. USA.* 93:196-200.

Underhill, P.A. Jin, L., Lin, A.A., Mehdi, S.Q., Jenkins, T., Vollrath, D., Davis. R.W. et al. 1997. Detection of numerous Y chromosome biallelic polymorphisms by denaturing high-performance liquid chromatography. *Genome Res.* 7:996-1005.

Underhill, P.A., Shen, P., Lin, A.A., Jin, L., Passarino, G., Yang, W.H., Kauffman, E. et al. 2000. Y chromosome sequence variation and the history of human populations. *Nature Genet.* 29:358-361.

Vigilant, L., Stoneking, M., Harpending, H., Hawkes, K. and Wilson, A. C. 1991. African populations and the evolution of human mitochondrial DNA. *Science* 253:1503-150.

Weng, Z., Yuan, Y. and Du, R. 1989. Analysis of the genetic structure of human populations in China. *Acta Anthropol. Sin.* 8:261-268.

Wilson, A.C. and Cann, R. 1992. The recent African genesis of humans. *Sci. Am.* 266:68-73.

Wolpoff, M.H., Hawks, J. and Caspari, R. 2000. Multiregional, not multiple origins. *Am. J. Phys. Anthropol.* 112:129-136.

Wu, X.Z. and Poirier, F.E. 1995. *Human Evolution in China.* Oxford: Oxford University Press.

Zhang, H. 1988. Distribution of dermatoglyphic parameters in fifty-two Chinese populations. *Acta Anthropol. Sin.* 7:39-45.

Zhang, Z.B. 1988. An analysis of the physical characteristics of modern Chinese. *Acta Anthropol. Sin.* 7:314-323.

Zhao, T., Zhang, G., Zhu, Y., Zheng, S., Liu., D., Chen, Q. and Zhang, X. 1986. The distribution of immunoglobulin Gm allotypes in forty Chinese populations. *Acta Anthropol. Sin.* 6:1-8 (in Chinese).

Zhao, T.M., and Lee, T.D. 1989. Gm and Km allotypes in 74 Chinese populations: A hypothesis of the origin of the Chinese nation. *Hum. Genet.* 83:101-110.

Zhao, Z. M., Jin, L., Fu, Y.X., Ramsay, M., Jenkins, T., Leskinen, E., Pamilo P. et al. 2000. Worldwide DNA sequence variation in a 10kb noncoding region on human chromosome 22. *Proc. Natl. Acad. Sci. USA* . 97: 11354-11358.

Part IV: The Peopling of Oceania

THE GENETIC TRAIL FROM SOUTHEAST ASIA TO THE PACIFIC

RANJAN DEKA
University of Cincinnati

BING SU
Kunming Institute of Zoology, Chinese Academy of Sciences
Human Genetics Center, University of Texas
School of Life Sciences, Fudan University

LI JIN
Human Genetics Center, University of Texas
School of Life Sciences, Fudan University

Southeast Asia is of outstanding anthropological interest because of its central position in important migratory events during the prehistoric past. Scholars in general agree with the events associated with the first population expansion leading to the colonization of Australia and New Guinea between 40,000 and 50,000 years ago, when Pleistocene hunter-gatherers crossed from Indonesia to the Sahul landmass (Bellwood 1997; Roberts et al. 1990). The second migration, which places Southeast Asia more prominently on the anthropological map of recent population movements, is connected with the peopling of the Polynesian islands. In this paper, we will examine the issues relating to this second event.

In spite of amassing a large body of archaeological, linguistic and more recent genetic data from the general area of interest, the question surrounding the colonization of Polynesia has nonetheless remained controversial. Two principal hypotheses have emerged out of the scientific investigations. The first, dubbed the 'express train to Polynesia' (Diamond 1988) and based on archaeological and linguistic evidence (Bellwood 1978), posits a very rapid eastward migration of Neolithic farmers from Southern China about 4,000 to 5,000 years before present, spreading Austronesian languages and the associated Lapita culture and culminating in the colonization of Polynesia. In this model, Taiwan, which is adjacent to the Asian mainland, was colonized first and the Austronesian expansion to Polynesia from there continued through the Philippines, Indonesia and Borneo; with the migrants interbreeding with or replacing the local

Australoid populations of the area (Bellwood 1997; Blust 1988). The process was completed in just about 2,000 years covering a distance of about 10,000 kilometers—this was perhaps the fastest human migration in prehistory! This interpretation of linguistic and archaeological data received strong support when mtDNA apparently showed a connection between the Taiwanese aborigines and the Polynesians (Sykes et al. 1995; Melton et al. 1995; Redd et al. 1995) placing the 'express train' on a firm footing. A second hypothesis proposed by Terrell (1988) refutes the above position and argues for a neighboring homeland for the Polynesians. This model, often referred to as the 'entangled bank' metaphor, asserts that the Polynesians evolved in a complex process of interactions through trade and admixture among the already settled Pacific populations. Although the controversy continues, most scholars have favored the 'express train' hypothesis over the 'entangled bank' model, which has much weaker support from genetic data. More recently, Gray and Jordan (2000) presented a parsimonious language tree based on 77 Austronesian languages from the area, the topology of which highly conforms to the express train model.

The Genetic Trail

Genetic evidence largely supports the basic tenet of the 'express train' hypothesis that the Polynesian founders came from the Southeast Asian agricultural heartland (Hill and Serjeantson 1989). The initial molecular studies using nuclear markers, including serum protein and blood groups had established the relationship between Southeast Asian and Polynesian populations (Kirk 1989). Globin gene analysis, although largely supporteding a Southeast Asian connection, did not completely rule out some Melanesian contribution (Hill et al. 1987; Boyce et al. 1995). Martinson (1996) dealt with this subject in great detail, especially accounting for the 30% Melanesian globin gene component in the Polynesians. He takes a less extreme position of the 'express train' model suggesting that although the Polynesian gene pool is largely derived from island Southeast Asia; on their way to Polynesia, the migrants had acquired some genes from the Melanesians. The strongest support for the Southeast Asia-Polynesia connection was, however, provided by more recent mtDNA evidence. The famous 9-base pair (bp) deletion, shown to be an informative marker for Southeast Asian populations (Wrischnik et al. 1987; Ballinger et al. 1992), was found in near complete fixation in Polynesia (Hertzberg et al. 1989;

Hagelberg and Clegg 1993; Lum et al. 1994) and absent in Papua New Guinea and Australian aborigines (Hertzberg et al. 1989).

The mitochondrial DNA scenario: The Southeast Asian connection was established. The issue of contention was, however, whether the voyage to Polynesia had actually begun in Taiwan. The above studies did not explicitly point to Taiwan as the originating station. In the mean time, Hagelberg and Clegg (1993) reported a characteristic pattern of base substitutions in the control region of the mtDNA genome, which provided new insight into this question. Three substitutions at positions 16217, 16247 and 16261 on the 9-bp deletion background became the genetic hallmark of Polynesian ancestry (Melton et al. 1995; Redd et al. 1995). The ancestral type (Anderson sequence) at these positions is a "TAC" sequence (Fig 1). Melton et al. (1995) coined the term "Polynesian motif" to describe the characteristic substituted sequence "CGT", which was found in a very high frequency (≥ 80%) among the Polynesians and in the corridor through the Philippines and Indonesia. Related intermediate types, as shown in Figure 1, also were found in appreciable frequencies in this area of distribution and the highest diversity was found in the Taiwanese populations. Melton et al. (1995). proposed the following sequence of events in the evolution of the motif: the first substitution had occurred in the 9-bp deletion background at position 16217 resulting in the type "CAC", which was distributed throughout Southeast Asia, followed by a change at position 16261 on this background resulting in the type "CAT", which had probably occurred in Taiwan and diffused from here around 6,000 years ago. The final change had occurred at position 16247 in eastern Indonesia giving rise to the "CGT" motif. Based on these observations, the origin of the Polynesian motif was traced to Taiwan.

Figure 1. The Polynesian Motif

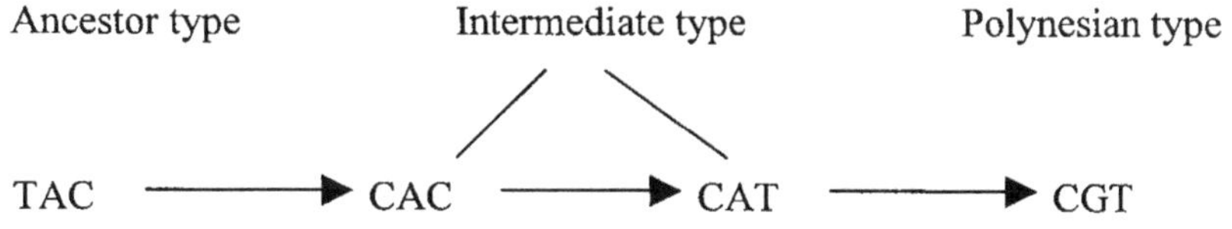

The above scheme of events appeared to provide unequivocal support to the express train hypothesis. However, there were questions that remained to be addressed. Are there other plausible scenarios or alternative explanations of the mtDNA data? A recent reanalysis by Richards et al. (1998) of the published mtDNA data (Sykes et al. 1995; Redd et al. 1995) questioned the validity of the above proposition. Richards et al. (1998) argued that, based on assessment of divergence times for the motif and age estimates of the relevant populations, mtDNA data do not support a Taiwanese origin. Rather, the evidence is more consistent with an island Southeast Asian ancestry with the homeland being in eastern Indonesia. Incidentally, Redd et al. (1995) and Melton et al. (1995) recognized that the full motif had evolved in eastern Indonesia.

Aside from Richard et al.'s (1998) explanation, there are other scenarios that ought to be considered with respect to mtDNA data. Although the "Polynesian motif" in its final form has a restricted and defined distribution, the other derived types are widely distributed all over the greater Asian domain, including Mongolia (Melton et al. 1995; Kolman et al. 1996). Further we found that the "CAT' motif, which seemed to trace an origin to Taiwan, has in fact a much wider distribution throughout Southeast Asia, including southern China. Importantly, we also found that recurrent mutation encompassing the motif (Fig. 2) is rather high (data not shown) diluting its effectiveness and obscuring the track of Polynesian migrations.

Figure 2. The Polynesian Motif & Recurrent Mutations

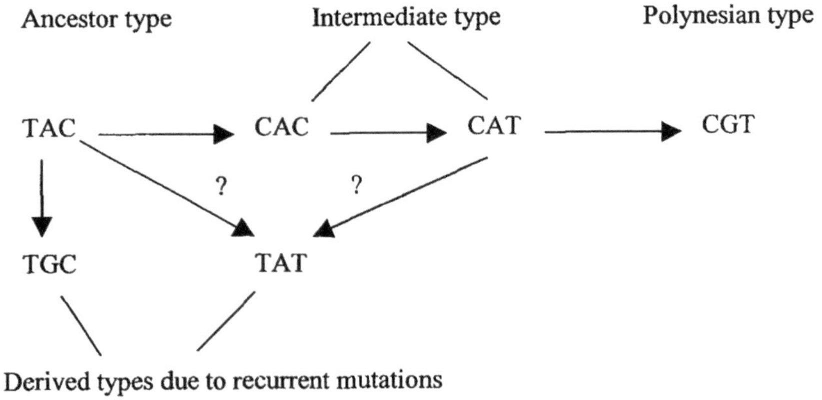

The Y chromosome evidence: The Y chromosome is similar to the mitochondrial DNA genome because it also escapes recombination during meiosis, and therefore, retains an intact evolutionary history. In recent years, the power of Y-chromosome markers has been greatly recognized in tracing human history and migration (Hammer 1995; Jobling & Smith 1996; Thomson et al. 2000). It becomes even more effective with the proper choice of markers. Neutral biallelic polymorphisms are especially ideal for evolutionary reconstruction for two simple but important reasons—their mutation rate is low and recurrent mutation at these sites is almost absent. Thus for evolutionary studies, biallelic markers located on the non-recombining portion of the Y chromosome are more powerful than mtDNA sequence information, which is subject to higher mutation rates and recurrent mutation. Recently, we published a report on the origin of Polynesian people based on an analysis of 19 biallelic Y-chromosome markers (Su et al. 2000). This study was based on more than 550 males drawn from 36 populations living in greater Southeast Asia, Micronesia, Melanesia and Polynesia. In all, we identified 15 haplotypes based on polymorphisms at the 19 loci. This data is summarized in Table 1, and detailed haplotype distribution in specific populations is given in Su et. al. (2000).

The distribution of the Y-chromosome haplotypes shows a characteristic pattern. We had earlier shown that H1 is the ancestral haplotype based on its appearance in chimpanzees, H2 is also a relatively ancient haplotype because of its appearance in both African and non-African populations, and H5 is the common ancestor of all non-African haplotypes (Su et al. 1999). Of the 15 observed haplotypes, 14 are present in the Southeast Asian populations, who with a haplotypic diversity of 88% are by far the most diverse group among all of the studied populations. The rest of the geographic populations share a subset of these haplotypes—seven of them are present in the Taiwanese (diversity 70%) and 10 are shared by the Micronesians and Polynesians (diversity 72%). Melanesians share four haplotypes with a distinct one, H17, which is apparently a Melanesian-specific haplotype with its presence only in two Micronesian individuals. The most noteworthy feature of the Y haplotype distribution, however, is a striking difference between the Taiwanese and Micronesian/Polynesian populations. With the exception of H6, these two groups of populations have two independent sets of haplotypes. H1, H2, H4 and H5 are exclusively present among the Micronesian/Polynesian groups. On the other hand, H7, H8, H9, H10, H11 and H12 are found only in the Taiwanese populations, with occasional presence in Micronesia. Conspicuously, both the ancestral (H1) and the non-African ancestral (H5) haplotypes are absent in Taiwan.

Table 1. Distribution of Y-chromosome haplotypes in Southeast Asia and Oceania

	Geographic Populations (N)				
Haplotype	SE Asia (267)	Taiwan (58)	Melanesia (113)	Micronesia (73)	Polynesia (40)
H1	7.1		14.2	8.2	42.5
H2	1.9			1.4	
H3	2.6				
H4	3.4		1.8	2.7	
H5	11.6		45.1	49.3	22.5
H6	11.6	12.1	2.6	24.7	32.5
H7	4.5	1.7			
H8	11.6	5.2		2.7	
H9	10.1	48.3			
H10	3.8	18.9		2.7	
H11	23.6	10.3			
H12	6.4	3.4		4.1	
H14	1.5			1.4	2.5
H16	0.4				
H17			36.3	2.7	

Evidently, the Taiwanese and the Micronesian/Polynesian populations have very little in common on the Y chromosome. This would not be the case had Polynesian ancestry been derived from Taiwan. Does the Y chromosome evidence completely dispute the express train model? Not necessarily! What we see here are two independent sets of haplotypes in the two locales, both of which are, however, widely distributed throughout the greater Southeast Asian continent including southern China, mainland and insular Southeast Asia. Linguistically, the Southeast Aisan population groups are diverse and include Sino-Tibetan, Hmong Mien, Austroasiatic and Austronesian speakers. In our PNAS paper (Su et al. 2000), we have shown that the haplotypes H1 through H14 are randomly distributed among all of these populations, without any discernible affiliation. This strongly suggests a genetic continuum throughout greater Southeast Asia. Anecdotally, our autosomal microsatellite data (Parra et al. 1999) also suggest a similar continuum. We had studied nine autosomal microsatellite loci in 16 Asian and Pacific populations. Although the number of loci is limited, the neighbor-joining tree did not show any clustering of populations based on linguistic or geographic affiliation (Fig. 3). This perhaps reflects a recent common origin of these populations as noticed from the Y chromosome data.

The occurrence of both Taiwanese and Polynesian haplotypes in Southeast Asia together with the highest haplotypic diversity, supports the notion of a Southeast Asian homeland for both of these population groups. However, the data do not show a connection between the Taiwanese and the Polynesians. The most likely explanation of the Y chromosome data is that both the Taiwanese and the Polynesians derive their ancestry in Southeast Asia. However, colonization of Polynesia had occurred via a route independent of the expansion toward Taiwan. The random distribution of haplotypes, together with the genetic continuity in Southeast Asia obscures the actual center of origin of the Polynesians. These observations do not refute the express train model in its entirety. It is plausible that the Polynesian colonists had taken a route through island Southeast Asia, a view supported by Richard et al.'s (1998) reanalysis of the mtDNA data.

The extent of the Melanesian contribution to the colonization of Polynesia is debatable (Terrell 1988; Lum et al. 1998). The Y-chromosome haplotype H17 is almost exclusively restricted to Melanesia, with a very low frequency in Micronesia (see Table 1). It is unlikely that there was any significant male-mediated Melanesian contribution to the settlement of Polynesia. However, a comparatively higher proportion of Melanesian alleles at the nuclear and mtDNA genomes has been found in Polynesia (Boyce et al. 1995; Sykes et al. 1995). At this time, the reasons for these incongruities are not clear. However, it should be noted that the geographic dispersal of human populations has been shown to bear signatures of sex-dependent migration (Seielstad et al. 1998; Pérez-Lezaun et al. 1999), which could have played important roles in the Pacific colonization as well (Lum et al. 1998).

Figure 3. Neighbor-joining tree for 16 Asian populations based on D_{SW} distance. ***Austronesians:*** Atayal, Ami (Taiwan), Malay (Malaysia), Javanese (Java), Batak (Sumatra), Filipino (Philippines), Samoan (Samoa); ***Tibetoburman:*** Kachari (Northeast India), Japanese, Sinitic: Chinese; ***Daic:*** N-Thai, NE-Thai (Thailand); ***Austroasiatic:*** Cambodian (Kampuchea), So (Thailand), Orang Asli (Malaysia); ***Papuan:*** New Guinea Highlander.

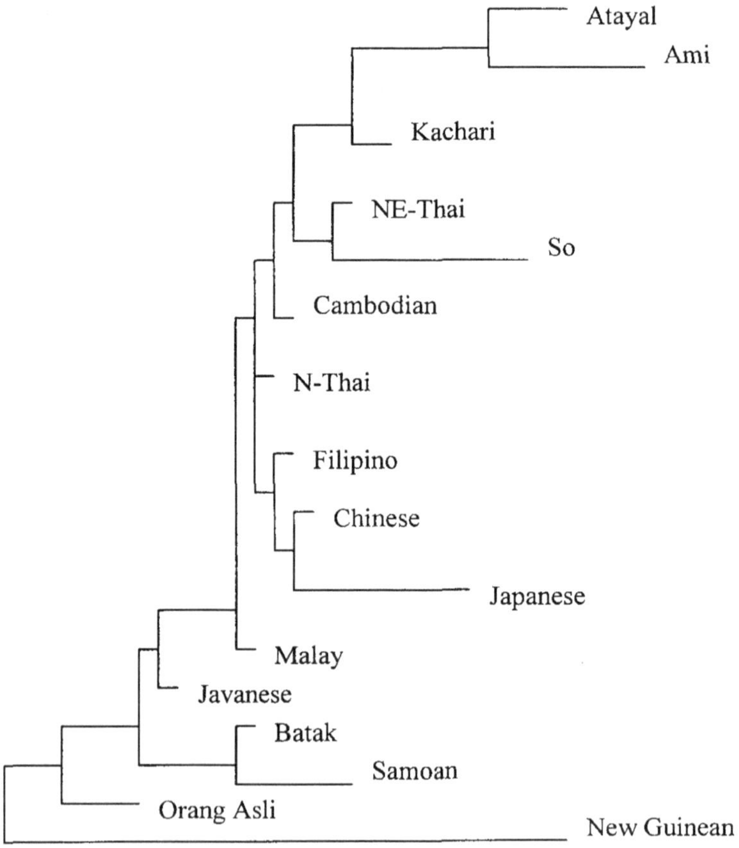

Current evidence that Southeast Asia holds the ancestral position in the cascade of Polynesian settlement is strong. Nonetheless, there are several important issues which need to be addressed for a complete understanding of the entire process of these prehistoric population expansions. More rigorous, thorough and comprehensive analysis of genetic diversity in Southeast Asia will shed important light in the study of prehistoric movements in this anthropologically and evolutionarily important region of the world. Cavalli-Sforza et al. (1994) had remarked that genetic data for Southeast Asia is too sparse to make compelling conclusions regarding the genetic relationships of the populations of the area!

Acknowledgements

We thank many colleagues who have contributed to the Y-chromosome study described here. Among them, Drs. Jeremy Martinson, Peter Underhill, Stephen T. McGarvey and Ranajit Chakraborty deserve our special mention. We were supported by grants from the National Institutes of Health and the National Science Foundation, USA and the Chinese National Natural Science Foundation. LJ was supported by the Li Foundation and RD was supported by the Center for Environmental Genetics, University of Cincinnati.

References

Ballinger, S.W., Schurr, T.G., Torroni, A., Gan, Y.Y., Hodge, J.A., Hassan, K., Chen, K-H. and Wallace, D.C. 1992. Southeast Asian mitochondrial DNA analysis reveals genetic continuity of ancient mongoloid migrations. *Genetics* 130:139-152.

Bellwood, P. 1978. Man's conquest of the Pacific: The Prehistory of Southeast Asia and Oceania. New York: Oxford University Press.

Bellwood, P. 1997. *Prehistory of the Indo-Malaysian Archipelago*. 2nd Edition. Honolulu: University of Hawaii Press.

Blust, R. 1988. The Austronesian homeland: a linguistic perspective. *Asian Perspect.* 26:45-67.

Boyce, A.J., Harding, R.M. and Martinson, J.J. 1995. Population genetics of the α-globin gene complex in Oceania. IN: Boyce, A.J. and Reynolds, V.R. (eds.) *Human Populations: Diversity and Adaptation*, Oxford: Oxford University Press, pp. 217-232.

Cavalli-Sforza, L.L., Menozzi, P., and Piazza, A. 1994. *The History and Geography of Human Genes*. Princeton: Princeton University Press.
Diamond, J.M. 1988. Express train to Polynesia. *Nature* 336:307-308.
Hammer, M.F. 1995. A recent common ancestry for human Y chromsomes. *Nature* 378: 376-378.
Gray, R.D. and Jordan, F.M. 2000. Language trees support the express-train sequence of Austronesian expansion. *Nature* 405:1052-1055.
Hagelberg, E. and Clegg, J.B. 1993. Genetic polymorphism in prehistoric Pacific islanders determined by analysis of ancient bone DNA. *Proc. Royal Soc. London* 252:163-170.
Hertzberg, M.S., Mickleson, K.N.P., Serjeantson, S.W., Prior, J.F. and Trent, R.J. 1989. An Asian-specific 9-bp deletion of mitochondrial DNA is frequently found in Polynesians. *Am. J. Hum. Genet.* 44:504-510.
Hill, A.V.S., Gentile, B., Bonnardot, J.M., Roux, J., Weatherall, D.J. and Clegg, J.B. 1987. Polyneisan origins and affinities: globin gene variants in eastern Polynesia. *Am. J. Hum. Genet.* 40:453-463.
Hill, A.V.S. and Serjeantson, S.W. (eds). 1989. *The Colonization of the Pacific: A Genetic Trail*. Oxford: Clarendon.
Jobling, M.A. and Tyler-Smith, C. 1995. Fathers and sons: the Y chromosome and human evolution. *Trends Genet.* 11:449-455.
Kirk, R.L. 1989. Population genetic studies in the Pacific. IN: Hill, A.V.S. and Serjeantson, S.W. (eds.), *The Colonization of the Pacific: A Genetic Trail*. Oxford: Oxford University Press, pp. 60-119.
Kolman, C., Sambuughin, N. and Bermingham, E. 1996. Mitochondrial DNA analysis of Mongolian populations and implications for the origin of New World founders. *Genetics* 142:1321-1334.
Lum, J.K., Rickards, O., Ching, C. and Cann, R.L. 1994. Polynesian mitochondrial DNAs reveal three deep maternal lineage clusters. *Hum. Biol.* 66:567-590.
Lum, J.K., Cann, R.L., Martinson, J.J. and Jorde, L.B. 1998. Mitochondrial and nuclear genetic relationships among the Pacific island and Asian populations. *Am. J. Hum. Genet.* 63:613-624.
Martinson, J.J. 1996. Molecular perspectives on the colonisation of the Pacific. IN: Boyce, A.J. and Mascie-Taylor, C.G.N. (eds.) *Molecular Biology and Human Diversity*. Cambridge: Cambridge University Press, pp 171-195.
Melton, T., Peterson, R., Redd, A.J., Saha, N., Sofro, A.S.M., Martinson, J. and Stoneking, M. 1995. Polynesian genetic affinities with Southeast Asian populations as identified by mtDNA analysis. *Am. J. Hum. Genet.* 57:403-414.
Parra, E., Saha, N., Soemantri, A.G., McGarvey, S.T., Hundrieser, J., Shriver, M.D., and Deka, R. 1999. Genetic variation at 9 autosomal microsatellite loci in Asian and Pacific populations. *Hum. Biol.* 71:757-779.
Pérez-Lezaun, A., Calafell, F., Comas, D., Mateu, E., Bosch, E., Martínez-Arias, R., Clarimón, J., et al. 1999. Sex-specific migration patterns in Central Asian populations revealed by analysis of short tandem repeats and mtDNA. *Am. J. Hum. Genet.* 65:208-219.

Redd, A., Takezaki, N., Sherry, S., McGarvey, S., Sofro, A.S.M. and Stoneking, M. 1995. Evolutionary history of the COII/tRNA^{Lys} intergenic 9-bp deletion in human mitochondrial DNAs from the Pacific. *Mol. Biol. Evol.* 12:604-615.

Richards, M., Oppenheimer, S. and Sykes, B. 1998. MtDNA suggests Polynesian origins in Eastern Indonesia. *Am. J. Hum. Genet.* 63:1234-1237.

Roberts, R.G., Jones, R. and Smith, M.A. 1990. Thermoluminescence dating of a 50,000-year-old human occupation site in northern Australia. *Nature* 345:153-156.

Seielstad, M.T., Minch, E. and Cavalli-Sforza, L.L. 1998. Genetic evidence for a higher female migration rate in humans. *Nature Genet.* 20:278-280.

Su, B., Xiao, J., Underhill, P., Deka, R., Zhang, W., Akey, J., Huang, W., et al. 1999. Y-chromosome evidence for a northward migration of modern humans into Eastern Asia during the last ice age. *Am. J. Hum. Genet.* 65:1718-1724.

Su, B., Jin, L., Underhill, P., Martinson, M., Saha, N., McGarvey, S.T., Shriver, M.D., et al. 2000. Polynesian origins: Insights from the Y chromosome. *Proc. Natl. Acad. Sci. USA* 97:8225-8228.

Sykes, B., Leiboff, A., Low-Beer, J., Tetzner, S. and Richards, M. 1995. The origins of the Polynesians: an interpretation from mitochondrial lineage analysis. *Am. J. Hum. Genet.* 57:1463-1475.

Terrell, J. 1988. History as a family tree, history as an entangled bank: constructing images and interpretations of prehistory in the South pacific. *Antiquity* 62:642-657.

Thomson, R., Pritchard, J.K., Shen, P., Oefner, P.J. and Feldman, M.W. 2000. Recent common ancestry of human Y chromsomes: evidence from DNA sequence data. *Proc. Natl. Acad. Sci. USA* 97:7360-7365.

Wrischnik, L.A., Higuchi, R.G., Stoneking, M., Arnheim, N. and Wilson, A.C. 1987. Length mutations in human mitochondrial DNA: direct sequencing of enzymatically amplified DNA. *Nucleic Acids Res.* 15:529-542.

THE COLONIZATION OF REMOTE OCEANIA AND THE DROWNING OF SUNDALAND

J. K. LUM

Tokyo Women's Medical University

Ever since European explorers ventured into the Pacific and realized that virtually every island had already been discovered and that most were still inhabited, Pacific Island settlement has piqued the interest of scholars and laymen alike. How were these, often miniscule, islands sprinkled across the largest ocean on the planet located and successfully colonized? Historically the Pacific Islands have been divided into three major groups, Melanesia, Micronesia, and Polynesia. These categories are based on both geography and the morphology of the islanders. Melanesia (black islands) includes islands in the Western Pacific where the people tend to have dark skin and kinky hair. In contrast, the people of Micronesia and Polynesia tend to have lighter skin and more Asiatic features. Micronesia (small islands) includes the relatively small islands of the North Pacific, while Polynesia (many islands) consists of the remaining Eastern Pacific archipelagos within the triangle bounded by Hawaii, Rapanui (Easter Island), and Aotearoa (New Zealand) (Figure 1). As more information describing the cultures, languages, physical characteristics, and more recently genetics has accumulated, it has become clear that these groupings are inconsistent with many aspects of human diversity (e.g. the Polynesian Outliers). In addition, over the past decade there has been a paradigm shift in Oceanic studies resulting from a growing acceptance of the feasibility of long distance voyaging among archipelagos (Lewis 1972; Irwin 1992; Finney 1994). Technological feasibility is however, only one aspect of contact. Ethnographic records often describe voyagers being met with hostility as locals sought to guard their limited island resources. Many examples of sustained contact among linguistically and culturally distinct island populations are however, also documented. These data raise a number of as yet unresolved questions. Under which circumstances are foreign migrants welcomed and long term interaction among populations sustained? When interacting networks of populations do develop, how is interaction controlled and how does this affect the evolution of gene pools, cultures, and languages? Concomitant with the questioning of isolation as the rule for the development of island societies has been an analogous questioning of a single origin, both geographic and temporal, of Pacific island populations (Terrell and Welsch

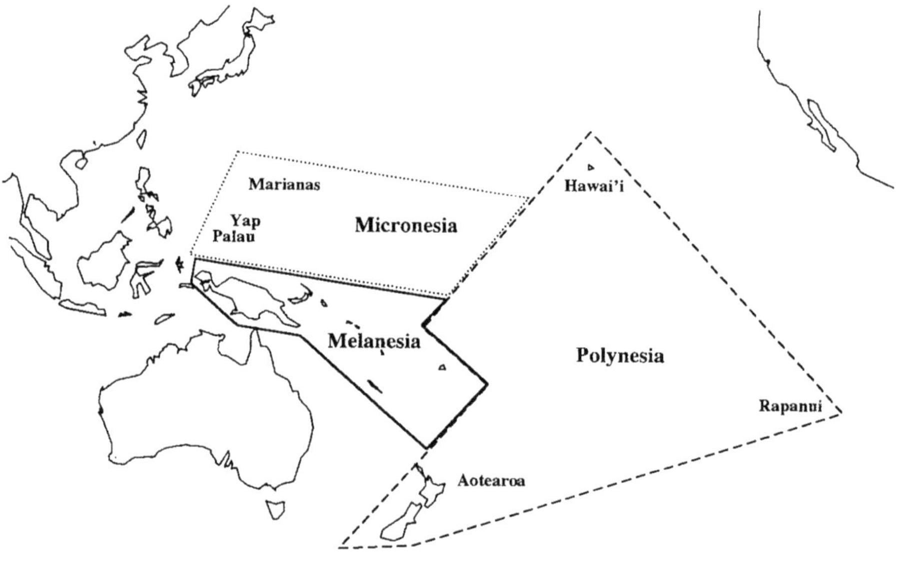

Figure 1. Map of the Pacific.

1997). Long standing hypotheses of interacting voyaging communities as the proximate source of Remote Oceanic populations (Terrell 1988; Irwin 1992) are currently being re-examined and new motivations for colonization have recently been proposed (Oppenheimer 1998). The following will summarize archaeological, linguistic, and morphological data from the Pacific, discuss some current ideas of settlement and interaction, and reconcile these with analyses of mitochondrial DNA (mtDNA) and briefly, other genetic markers.

Archaeology

The settlement of the Pacific appears as two punctuated expansions in the archaeological record. The earliest evidence of human colonization of the Pacific is found in New Guinea dated to 40,000 years ago (Groube et al. 1989). Colonization of nearby islands continued eastward and finally reached the Northern Solomon Islands by 29,000 years ago (Wickler and Spriggs 1988). It remains unclear how much contact these Pleistocene colonists maintained with each other or with ancestral populations further West. It is clear, however, that these populations independently developed root crop-based agriculture in the New Guinea highlands perhaps as early as 9,000 years ago (Golson 1977; White and O'Connell 1982), predating the grain-based agriculture of Asia.

The second wave of colonization occurred within the last 4,000 years and included all of Eastern Melanesia, Micronesia, and Polynesia. This Holocene expansion is associated with the Lapita cultural complex in Melanesia and Western Polynesia (Kirch 1997). The emblem of this cultural complex is highly stylized, dentate stamped, lime in-filled, red slipped pottery. This distinctive pottery has been found on islands off the north coast of New Guinea, the Southern Solomons, New Caledonia, Vanuatu, Fiji, and Samoa in what has been called an archaeologically instantaneous horizon dated to 3,500 years before present (Kirch and Hunt 1988). Similar pottery, particularly "Lapitiod" pottery and Marianas Redware have also been unearthed in Micronesia (Spriggs 1989; Intoh, 1997). Other components of the Lapita cultural complex found throughout Melanesia and the high islands of Polynesia include the pig, dog, and chicken. Although only the dog was found on Pohnpei in Central Micronesia, all three are found in the lowest stratum of Fais, one of the outer islands of Yap (Intoh, 1997).

Language

All of the languages of the Pacific are grouped into either the Austronesian or Papuan language families. The Austronesian languages of the Pacific are, with two or three exceptions, grouped within the Oceanic branch of the language family (Pawley and Green 1973). The Oceanic languages are considered a monophyletic group within a well defined Austronesian language family which also includes languages found throughout Island Southeast Asia, Taiwan, parts of mainland Southeast Asia, and also Madagascar. The two non-Oceanic Austronesian languages of the Pacific are found in the Western Micronesian archipelagos of Palau and the Marianas. Both Palauan and Chammoro are considered linguistic isolates and are only distantly related to each other or to languages of Island Southeast Asia. Previously Yapese, the third language of Western Micronesia was also considered an isolate (Bender 1971), but Ross (1996) has argued that although it contains extensive non-Oceanic borrowing, most likely from Palauan, it should be considered an Oceanic language. The linguistic homogeneity of Austronesian languages is in sharp contrast to the Papuan languages.

The Papuan languages are extremely diverse and were originally defined by default as non-Austronesian languages of the Pacific. Papuan languages are concentrated in inland New Guinea and the highlands of the Northern Solomon Islands (Wurm and Hattori 1981). Thus their distribution is coincident with regions of Pleistocene settlement. Furthermore, the diversity of Papuan languages relative to the Austronesian languages mirrors the difference in age of the Pleistocene and Holocene waves of Pacific Island settlement. Based upon these archaeological, linguistic, and also coincident biogeographical patterns, the recently settled regions inhabited by Austronesian-speaking populations have been grouped into a single bio-cultural unit called Remote Oceania (Pawley and Green 1973; Green 1991; Green 1999).

Morphology

As mentioned previously, the definition of Melanesia is based in part on morphological characteristics. In general, Melanesians have dark skin and kinky hair relative to Micronesians, Polynesians, and Asians. This difference is more than skin deep as indicated by craniometric analyses (Pietrusewsky 1990a; Pietrusewsky 1990b) which cluster both Papuan and Austronesian-speaking Melanesians from Pleistocene and Holocene settled regions with aboriginal Australians to the exclusion of Polynesians, Asians, and most Micronesians. Thus, the morphological patterns observed are inconsistent with the archaeological and linguistic patterns described above. This inconsistency raises a number of questions. Why are the Austronesian-speaking Melanesians morphologically more similar to Papuan-speaking Melanesians than to other Remote Oceanic Islanders? If this similarity is the result of gene flow among Melanesians, then did this gene flow occur within the last 3,500 years, or before the settlement of Remote Oceania? Alternatively, is this seeming inconsistency the result of the misclassification of Melanesian Austronesian languages?

The Colonization of Remote Oceania

Numerous scenarios to explain the settlement of the Pacific with respect to the data described above have been proposed. Of these I will focus this discussion on three sets of ideas. Two of these represent extreme views of the proximate origin of the Remote Oceanic islanders and the extent of interactions during the Pacific colonization. The "Express Train" (Diamond 1988) specifies an intrusive colonization of Remote Oceania by a migration extending from Southern China. In contrast, the "Entangled Bank" (Terrell 1988) views the expansion into Remote Ocenaia as an outgrowth of interacting Western Pacific populations. The third, "Eden in the East" (Oppenheimer 1998) is a recently proposed intermediate scenario incorporating sea level rises at the end of the Pleistocene as a motivation for the development of voyaging technology and ultimately, the colonization of Remote Oceania.

The Express Train

Diamond (1988) coined the phrase "Express Train to Polynesia" to describe what he saw as a rapid expansion from Southern China into Polynesia. His view has both archaeological and linguistic components. The archaeological basis for the hypothesis has two main parts. First, there is the rapid colonization of Remote Oceania after more than 25,000 years of Western Melanesian settlement. Second, this expansion is associated with a novel archaeological horizon in Island Southeast Asia (Bellwood 1985) and the archaeologically instantaneous appearance of the Lapita cultural complex extending from the Bismark Archipelago, through Eastern Melanesia, and into Western Polynesia (Kirch and Hunt 1988). The rapid expansion into previously uninhabited islands associated with novel material artifacts suggested to Diamond the arrival of an intrusive population. The origin of this invasive group was suggested by comparative linguistics.

Since all of the Remote Oceanic populations speak Austronesian languages, the source of the newcomers was reasoned to come from the source of the Austronesian languages. All Austronesian languages are grouped into between two and five main branches. All of these save one are found only in Taiwan, while the last includes all the remaining languages dispersed from Rapanui to Madagascar. This pattern of diversity is interpreted by most linguists as an origin of Austronesian languages in Taiwan (Bellwood 1985; Blust 1995; Diamond 2000). The "Express Train" thus sees the proximate origin of Remote Oceanic islanders in Taiwan, or a population of Austronesian speakers on the Chinese mainland that has since disappeared. The Austronesian speakers are thought to have expanded from China 4-5,000 years ago (Blust 1981, cited in Bellwood 1985), spread through Island Southeast Asia rapidly replacing the indigenous languages, and finally out into Western Polynesia by 3,500 years ago. Although this view is compatible with the morphological similarity of Asians, Polynesians, and Micronesians, it leaves unexplained the question of the morphological similarity among Papuan and Austronesian-speaking Melanesians. It is also difficult to imagine how a complete linguistic replacement across this vast region that includes the large, rugged islands of Borneo and Sumatra could have occurred over such a short period of time (500-1,500 years). Consider that the "Express Train" next reached New Guinea where Austronesian languages are largely restricted to the north coast and off shore islands.

The Entangled Bank

The "Entangled Bank" is a metaphor originally used by Darwin (1859) to describe the intricate web of interacting selective forces contributing to the fitness of an organism that are difficult to discern by examining a single trait or characteristic. Terrell (1988) borrowed this analogy to describe his dissatisfaction with what he believes is an overly simplistic way of thinking about Pacific prehistory and the tyranny of linguistics. Terrell argues that Island Southeast Asian and Western Melanesian populations have been continuously interacting since the Pleistocene and it is this fertile exchange of ideas among cultures which spawned the rapid expansion into Remote Oceania without the need for an intrusive Asian migration. Furthermore, Terrell has pointed to the discordance between language and culture along the north coast of New Guinea (Welsch et al. 1992; but see also Moore and Romney 1994) to argue that linguistic patterns do not reflect Pacific settlement. Moreover, Terrell et al. (2001) argue that the Oceanic languages of Melanesia are hybrids of Papuan and Austronesian languages and so highlight extensive interaction among diverse Western Pacific populations rather than indicate an expansion of a defined linguistic group.

Although Terrell should be commended for championing the possibility of interaction among cultures and languages, he has, unfortunately been more specific about what he does not agree with than explicit about what he believes has occurred (Terrell 1988; Terrell et al. 1997; Terrell et al. 2001). In particular, he argues against the isolation of archipelagos without ever specifying which groups of people he believes were in contact or to what extent and in what ways they exchanged ideas or genes. This vagueness has detracted from the potential utility of his ideas. Remote Oceanic settlement by interacting Western Pacific populations is generally consistent with the observed morphological similarity among Papuan and Austronesian speaking Melanesians, but is inconsistent with the morphological similarity among Asians, Polynesians, and Micronesians. Terrell (1988) has suggested that the Polynesian morphology could result from a population bottleneck from Eastern Melanesia, but this unlikely event would have had to also occur in parallel during the settlement of Micronesia. Furthermore, such bottlenecks are contrary to Terrell's vague assertions of continuous interaction among populations.

Eden in the East

In his recent book "Eden in the East", Oppenheimer (1998) marshals data from a wide range of scientific disciplines and worldwide mythology to reconstruct a novel scenario of world prehistory. The 120 foot sea level rise at the end of the Pleistocene occurred in three rapid, discreet floods at 14,000, 12,000, and 8,000 years ago drowning the Sunda continental shelf and forming the Indonesian archipelago. The sea continued to rise until 6,000 years ago and then began to subside to the present level. Oppenheimer asserts that these catastrophic floods resulted in successive dispersals of people and ideas from Southeast Asia in all directions sparking the rise of Western Civilization in the Mediterranean and the settlement of Remote Oceania. I will focus here on his views pertaining to the latter.

One important contrast to the previously described ideas of Pacific colonization Oppenheimer presents concerns the origin of Austronesian languages. As mentioned previously, all but one of the deepest branches of the Austronesian language family are found exclusively in Taiwan while the other branch includes all other Austronesian languages and is absent from Taiwan. The higher diversity has led to the prevailing view among linguists that Taiwan is the source of the Austronesian language family (Blust 1995). Oppenheimer views this pattern of diversity in a different way. He argues that the linguistic diversity in Taiwan does not reflect age, but isolation. In his scenario, the Austronesian languages originated in interacting, coastal, maritime communities of Sundaland. The successive floods forced the Austronesian speakers to further refine their maritime technology and to form interacting networks in order to survive. He argues that the high degree of interaction among these communities maintained structural elements and vocabulary among most languages, but that Taiwan, being isolated at the extreme northern periphery was left out of the exchange. This view of the origin of Austronesian languages in Sundaland is appealing for several reasons. First, unlike the "Express Train" model, it does not require a rapid replacement of all the region's languages. It also allows for perhaps 14,000 or more years for the diversification of the Austronesian family rather than 4-5,000 years (Bellwood 1985).

Oppenheimer sees the settlement of Remote Oceania as a parallel expansion by two sets of competing populations. The first set of voyagers was a diverse group of populations from the Western Pacific that settled the main islands of Melanesia. These people were a mix of Island Southeast Asian and Western Melanesian populations who spoke Austronesian languages of the former, but had the general morphology of the latter. The second group of settlers were from what is now Eastern Indonesia and trailed the first group through Melanesia, settling the more remote, smaller islands (Polynesian Outliers) until branching out into Polynesia and Micronesia where the islands are generally smaller and ecologically similar to the ones they had become adapted. These people had the Asiatic morphology still found in Polynesia and Micronesia. Although the "Eden in the East" hypothesis is consistent with morphological patterns and the linguistic scenario is plausible, there are some difficulties. For example, linguistic analyses suggest that the Polynesian Outliers were recently settled from Western Polynesia (Pawley and Green 1973). If this is true, they cannot be the stepping stones to Polynesia Oppenheimer envisions. Although the archaeology and biology of most Polynesian Outliers have not been systematically examined, some appear to have been settled for at least 3,000 years (Irwin 1992) and so future studies may resolve this inconsistency.

mtDNA Analyses

mtDNA from a number of Pacific Island populations have been examined at different resolutions. One widely studied segment of mtDNA is region V (Cann and Wilson 1983). The globally common and evolutionarily ancestral sequence of region V includes two copies of a nine base repeat (Horai et al. 1993). A number of length polymorphisms have been reported from East Asian and Pacific populations including deletions of one repeat, additions of a third repeat, and expansions of single repeats (Wrischnik et al. 1987; Hertzberg et al. 1989; Lum and Cann 1998; Handoko et al. 2001). Initially, deleted individuals were only found in Asian and presumably Asian derived populations and so this length polymorphism became known as the "Asian deletion". Further studies have identified a number of independent deletions in Africa (Redd et al. 1995; Soodyall et al. 1996), Europe (Torroni et al. 1995), Southern India (Watkins et al. 1999), and Asia (Ballinger et al. 1992; Lum and Cann 1998). Notwithstanding, virtually all East Asians and all Pacific Islanders with the deletion form a monophyletic group (Lum and Cann 1998; Lum and Cann 2000). The region V deletion is found near fixation in Polynesia and in high frequency throughout Micronesia, with the exception of the Marianas (12%) (Hertzberg et al. 1989; Sykes et al. 1995; Lum and Cann 1998). Since the deletion is also found throughtout Asia, but is absent in the Highlands of New Guinea, these data indicate an Asian origin of most Polynesians and Micronesians.

To further refine the variability within the deletion cluster, control region sequences associated with the region V deletion have also been examined.

Table 1 shows the 48 sequences of 624 region V deleted individuals from 21 populations: Asia (3), Island Southeast Asia (3), Micronesia (9), and Polynesia (6) (Vigilant et al. 1991; Torroni et al. 1993; Lum et al. 1994; Sykes et al. 1995; Redd et al. 1995; Lum et al. 1998; Lum and Cann 2000). The lineages of Asian and Pacific populations are divided into five groups based upon length polymorphisms in region V and single base control region substitutions (see below) (Lum and Cann 2000). Group I includes region V deleted sequences and is further divided into the five subgroups (I.1, I.2, I.3, I.4, and I.5) as shown in Table 1. Three of these lineages (L1, L22, and L36) represent the majority of individuals (436/624) and are related to each other by successive single base substitutions; L36 gave rise to L22 (substitution at base 16261; numbering from Anderson et al. 1981), which in turn gave rise to L1 (substitution at base 16247). L1, L22, and L36 further gave rise to the three nested subgroups (I.1, I.2, and I.3). Subgroups I.4 and I.5 are sister to I.3 and each other. This relationship is mirrored by the geographic distribution of the subgroups. The most derived subgroup (I.1) predominates in Polynesian and Micronesia and is also found in Island Southeast Asia, but absent in Taiwan and mainland Asia. Similarly, subgroup I.4 is found in Island Southeast Asia and Western Micronesia, but not in Central-Eastern Micronesia, Polynesia, Taiwan or mainland Asia. In contrast, subgroup I.5 is found throughout Asia, but not in Remote Oceania. The extensive sequence diversity within group I indicates an ancient origin and radiation that is supported by the presence of subgroup I.2 and I.3 sequences in Amerindian populations (Lum et al. 1994; Lorenz and Smith 1997; Rickards et al. 1999). This distribution indicates an Asian origin of lineage group I, the dispersal of subgroups I.2 and I.3 at the time the Americas were colonized, and the proximate origin of most Micronesians and Polynesians in Island Southeast Asia (Lum and Cann 1998; Richards et al. 1998; Lum and Cann 2000).

As mentioned above, mtDNA sequences from Asian and Pacific populations have been divided into five groups based on length polymorphisms in region V and control region substitutions (Lum and Cann 2000). Four of these groups (I, II, II, IV) are distinct and well defined while the fifth (V) is an intermediate, default group. The frequencies of these five groups generated from sequences of 944 individuals from nine populations and regions are shown in Table 2 (Vigilant et al. 1989; Vigilant et al. 1991; Torroni et al. 1993; Lum et al. 1994; Sykes et al. 1995; Redd et al. 1995; Lum et al. 1998; Lum and Cann 2000). Although all of the group I sequences from Taiwan are shown in Table 1, the frequencies of these

Table 1. Distributions of Region V deleted lineages within Populations[a]

Subgroup	Lineage	Substitutions[b]	CHN	VNM	TWN	BOR	JAV	PHL	MRN	YAP	PAL	KIR	KSR	MRS	NAR	OIY	PON	SAM	CKS	MRQ	AOT	HWN	RPN
I.1	1	**189-217-247-261**	-	-	-	-	-	-	**5**	**26**	**33**	**6**	**6**	**3**	**15**	**45**	**9**	**42**	**46**	**12**	**20**	**22**	**11**
	2	**189-217-247-261-287**	-	-	-	-	-	-	-	-	-	**2**	**12**	**12**	-	-	-	-	-	-	-	-	-
	3	**189-217-247-261-362**	-	-	-	-	**1**	-	-	-	-	-	-	-	-	-	**1**	**2**	-	-	-	-	-
	4	**189-217-247-261-291**	-	-	-	-	-	-	-	-	-	-	-	-	-	-	-	**1**	**1**	-	-	-	-
	5	**189-217-247**	-	-	-	-	-	-	-	-	-	-	-	-	-	-	-	-	**1**	-	-	**1**	-
	6	188-189-217-247-261	-	-	-	-	-	-	-	1	-	-	-	-	-	-	-	-	-	-	-	-	-
	7	189-217-218-247-261-287	-	-	-	-	-	-	-	-	-	-	1	-	-	-	-	-	-	-	-	-	-
	8	189-217-247-248-259-261	-	-	-	-	-	-	-	-	-	-	-	-	1	-	-	-	-	-	-	-	-
	9	189-217-247-248-261	-	-	-	-	-	-	-	-	-	-	-	-	6	-	-	-	-	-	-	-	-
	10	189-217-247-261-262	-	-	-	-	-	-	-	-	-	-	-	-	-	-	-	1	-	-	-	-	-
	11	189-217-247-261-263	-	-	-	-	-	-	-	-	-	-	-	-	-	-	-	2	-	-	-	-	-
	12	189-217-247-261-274	-	-	-	-	-	-	-	-	-	-	-	-	-	1	-	-	-	-	-	-	-
	13	189-217-247-261-274(tv)	-	-	-	-	-	-	-	-	-	-	-	-	-	-	-	1	-	-	-	-	-
	14	189-217-247-261-295	-	-	-	-	-	-	-	-	-	-	-	-	1	-	-	-	-	-	-	-	-
	15	189-217-247-261-319	-	-	-	-	-	-	-	-	-	-	-	-	-	26	-	-	-	-	-	-	-
	16	189-217-247-261-342	-	-	-	-	-	-	-	-	-	-	-	-	-	-	-	2	-	-	-	-	-
	17	189-217-247-261-354(tv)	-	-	-	-	-	-	-	-	-	-	-	-	-	-	-	1	-	-	-	-	-
	18	189-217-247-261-361	-	-	-	-	-	-	-	-	-	-	-	-	-	-	-	1	-	-	-	-	-
	19	189-247-261	-	-	-	-	-	-	-	-	-	-	-	-	-	-	-	-	-	-	-	1	-
	20	189-217-247-262	-	-	-	-	-	-	-	-	2	-	-	-	-	-	-	-	-	-	-	-	-
	21	189-217-247-294	-	-	-	-	-	-	-	-	-	-	-	-	-	-	-	1	-	-	-	-	-

Table 1 (continued)

Subgroup	Lineage	Substit-utions[b]	CHN	VNM	TWN	BOR	JAV	PHL	MRN	YAP	PAL	KIR	KSR	MRS	NAR	OIY	PON	SAM	CKS	MRQ	AOT	HWN	RPN
I.2	22	**189-217-261**	-	-	**3**	**1**	-	**3**	**1**	**12**	**11**	**9**	**3**	**12**	**2**	**29**	**2**	**11**	**10**	**1**	**2**	**3**	**1**
	23	**189-217-261-362**	-	-	-	-	-	-	-	-	-	-	-	**1**	-	-	-	**1**	-	-	-	-	-
	24	**189-217-261-311**	**1**	-	**3**	-	-	-	-	-	-	-	-	-	-	-	-	**2**	-	-	-	-	-
	25	189-217-261-319	-	-	-	-	-	-	-	-	1	-	-	-	-	-	-	-	-	-	-	-	-
	26	189-209-217-223-261	-	-	-	-	-	-	-	-	2	-	-	-	-	-	-	-	-	-	-	-	-
	27	189-213-217-261	-	-	-	-	-	-	-	-	-	-	-	-	-	-	-	2	-	-	-	-	-
	28	189-217-261-312	-	-	-	-	-	-	-	-	-	-	-	-	-	-	-	-	1	-	-	-	-
	29	189-217-261-301	-	-	-	-	-	-	-	-	-	-	-	1	-	-	-	-	-	-	-	-	-
	30	189-217-223-261	-	-	-	-	-	1	-	-	-	-	-	-	-	-	-	-	-	-	-	-	-
	31	189-213-217-261-292	1	-	-	-	-	-	-	-	-	-	-	-	-	-	-	-	-	-	-	-	-
	32	189-217-234-256-261	-	1	-	-	-	-	-	-	-	-	-	-	-	-	-	-	-	-	-	-	-
	33	189-217-261-272-324	-	-	2	-	-	-	-	-	-	-	-	-	-	-	-	-	-	-	-	-	-
	34	189-217-261-272-288-324	-	-	1	-	-	-	-	-	-	-	-	-	-	-	-	-	-	-	-	-	-
	35	189-217-261-324	-	-	3	-	-	-	-	-	-	-	-	-	-	-	-	-	-	-	-	-	-

Table 1 (continued)

Subgroup	Lineage	Substitutions[b]	CHN	VNM	TWN	BOR	JAV	PHL	MRN	YAP	PAL	KIR	KSR	MRS	NAR	OIY	PON	SAM	CKS	MRQ	AOT	HWN	RPN
I.3	36	**189-217**	**1**	-	**1**	**1**	-	**3**	-	-	**5**	**1**	-	-	-	**3**	**3**	**1**	-	-	-	-	-
	37	**189-217-278**	-	-	-	-	-	-	-	**5**	-	**1**	-	-	**2**	**17**	**1**	-	-	-	-	-	-
	38	**189-217-274-335**	-	**1**	**3**	-	**1**	-	-	-	-	-	-	-	-	-	-	-	-	-	-	-	-
	39	**189-217-309-354**	**1**	**1**	-	-	-	-	-	-	-	-	-	-	-	-	-	-	-	-	-	-	-
	40	189-217-289	-	-	-	-	-	-	-	-	13	-	-	-	-	-	-	-	-	-	-	-	-
	41	189-217-278-311	-	-	-	-	-	-	-	-	-	-	1	-	-	-	-	-	-	-	-	-	-
	42	189-217-235	-	-	-	1	-	-	-	-	-	-	-	-	-	-	-	-	-	-	-	-	-
	43	189-217-218	-	1	-	-	-	-	-	-	-	-	-	-	-	-	-	-	-	-	-	-	-
	44	189-213-217-242-292-301	1	-	-	-	-	-	-	-	-	-	-	-	-	-	-	-	-	-	-	-	-
	45	189-217-365	-	-	3	-	-	-	-	-	-	-	-	-	-	-	-	-	-	-	-	-	-
I.4	46	**189-243**	-	-	-	**1**	-	**3**	-	**6**	**8**	-	-	-	-	-	-	-	-	-	-	-	-
I.5	47	**189-266(tv)**	**1**	**3**	**1**	**2**	**2**	-	-	-	-	-	-	-	-	-	-	-	-	-	-	-	-
	48	189-212-266(tv)	-	-	-	-	1	-	-	-	-	-	-	-	-	-	-	-	-	-	-	-	-

[a] Populations are: CHN = China; VNM = Vietnam; TWN = Taiwan; BOR = Borneo; JAV = Java; PHL = Philippines; MRN = Marianas; YAP = Yap Proper; PAL = Palau; KIR = Kiribati; KSR = Kosrae; MRS = Marshalls; NAR = Nauru; OIY = Outer Islands of Yap; PON = Pohnpei; SAM = Samoa; CKS = Cooks; MRQ = Marquesas; AOT = Aotearoa; HAW = Hawaii; RPN = Rapanui.

[b] Substitutions are numbered from 16000 and are relative to the reference sequence (Anderson et al. 1981); tv = transversion, all others are transitions. Lineages in multiple populations are boldface.

sequences have been scaled in Table 2 to reflect the total population deletion frequency reported by Sykes et al. (1995) rather than just those chosen for sequencing. As previously described (Lum and Cann 2000), the lineage groups in Asia and Papua New Guinea are largely non-overlapping. The majority of the sequences from Papua New Guinea belong to groups II and IV which are absent from Asia. Group III is found in Island Southeast Asia, but not mainland Asia, Taiwan, or New Guinea. In contrast, all four of the well-defined lineage groups are found in either Micronesia or Polynesia. Although widely distributed throughout Central-Eastern Micronesia and Polynesia (Lum et al. 1994; Sykes et. al 1995; Lum and Cann 2000), the lineage groups characteristic of New Guinea are found in low frequency (< 10%), and were most likely acquired as voyagers expanded through Melanesia along the north coast of New Guinea (Lum et al. 1994; Lum and Cann 2000).

The three western Micronesian archipelagos have distinct lingeage group profiles, consistent with the distinct languages spoken there. Neither Yap nor the Marianas contain groups II or IV suggesting direct colonization from Island Southeast Asia. Furthermore, the differences of group I diversities and frequencies of groups I and V suggest that they were settled independently (Lum and Cann 2000). As mentioned earlier, the Marianas have the lowest frequency of group I sequences (12%) of any Micronesian population. They are also the only Micronesian archipelago to have cultivated rice prehistorically (Hunter-Anderson et al. 1995). Thus, on the basis of language, mtDNA, and agriculture, the Marianas are distinct from all other Micronesian populations. Although the frequency of group I of the Marianas is similar, or lower than mainland Asian populations, the subgroups (1.1 and I.2) are characteristic of Pacific Islanders. This pattern of low diversity, but derived subgroups in the Marianas suggests that group I may be the result of post-settlement gene flow from the outer islands of Yap consistent with early historical documents (Barratt 1988).

Yap was connected with the outer islands to the East in the *sawei* tribute network. This entailed regular visits by fleets of outer island canoes bearing prestige items (Lingenfelter 1975; Alkire 1978). This exchange has resulted in the establishment of at least one maternal clan from the outer islands (Alkire 1972) with a corresponding mtDNA lineage (L37) on Yap (Lum and Cann 2000). Palau is unlike other Micronesian populations in that nearly a third of its sequences are inferred to be from New Guinea. The Palauan gene pool contains a thorough mix of lineages characteristic of Island Southeast

Table 2. Distribution of lineage groups (percent)

	N	I.1	I.2	I.3	I.4	I.5	II	III	IV	V
Papua New Guinea	17	-	-	-	-	-	35.3	-	47.1	17.6
Asia	49	-	6.1	12.2	-	8.2	-	-	-	73.5
Taiwan[a]	31	-	21.8	12.7		1.8	-	-	-	63.6
Island Southeast Asia	69	1.4	7.2	8.7	5.8	7.2	-	2.9	-	66.7
Marianas	52	9.6	1.9	-	-	-	-	-	-	88.5
Yap Proper	69	39.1	17.4	7.2	8.7	-	-	-	-	27.5
Palau	131	26.7	10.7	13.7	6.1	-	-	1.5	31.3	9.9
Central-Eastern Micronesia	278	52.9	21.2	10.4	-	-	3.6	-	2.2	9.7
Polynesia	239	70.7	14.2	0.4	-	-	8.4	0.4	-	5.9

[a] Frequencies of Lineage 1 subgroups are scaled to reflect the toal data of Sykes et al. 1995.

Asia, Central-Eastern Micronesia, and New Guinea consistent with its location at the intersection of these regions. It is unclear when these lineage groups came together, but they may also be partly the result of prehistoric exchange networks. Like the outer islanders who sailed East to Yap to bring valuables, the Yapese sailed Southeast to quarry their stone money discs from the rock islands of Palau (Alkire 1965; Lingenfelter 1975). These huge coral coins took up to ten years to carve, and consequently the Yapese lived for extended periods of time as guests of the Palauans. It is possible that the similarities between Palau and Yap (I.1, I.2, I.3, and I.4) reflect gene flow accompanying this prehistoric flow of valuables. Such gene flow would be consistent with the linguistic borrowing described by Ross (1996). The colonization routes and interactions suggested by the lineage group distributions in Table 2 are diagramed in Figure 2.

So how do these data correspond to the three sets of ideas present earlier? The complex mix of lineages predicted by extensive interaction envisioned in the "Entangled Bank" is only found in Palau. The mtDNA lineages of other populations from Micronesia and Polynesia are either virtually (>90%) or completely Asian derived (Table 2). Thus, with the possible exception of Palau, the mtDNA data do not support an origin of Micronesians or Polynesians from an extensively interacting Western Pacific network. This does not mean that culturally distinct archipelagos have not been interacting however. As described above, the Western Micronesian populations have been interacting with each other and with Central-Eastern Micronesians, but the effect on their mtDNA distinctiveness appears to have been minor. The similarities of the region V deleted lineages among Central-Eastern Micronesia populations and among Polynesian populations suggest that most gene flow that has occurred has been within rather than between regions. Such short-range gene flow within Central-Eastern Micronesia has been implicated in the diversification of both languages (Marck 1986) and gene pools (Lum 1998).

Although the mtDNA data are consistent with an Asian origin of most Micronesians and Polynesians proposed by both the "Express Train" and "Eden in the East" scenarios, several of the expectations of the "Express Train" model are not supported. In particular, the "Express Train" supposedly left Taiwan 4-5,000 years ago (Bellwood 1985), swept through Island Southeast Asia and into Melanesia and Western Polynesia 3,500 years ago (Diamond 1988). This suggests that Taiwan should be relatively diverse genetically and ancestral to Island Southeast Asia. Furthermore, the rapid

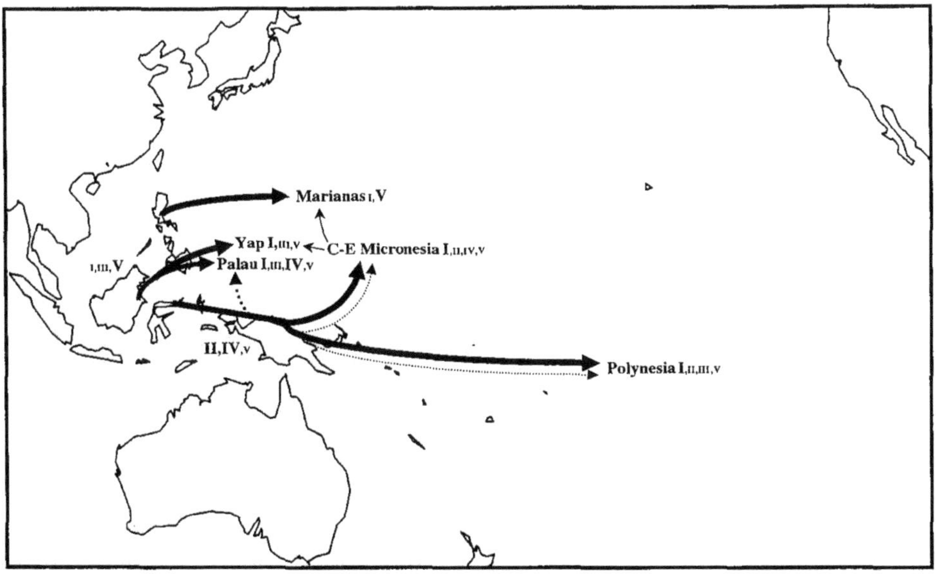

Figure 2. Settlement routes and post-settlement gene flow suggested by lineage group distributions (Table 2).

rate of expansion (500 to 1,500 years) suggests that the mtDNA lineages found in Island Southeast Asia and Remote Oceania should be subsets of those in Taiwan. This genetic pattern is not observed. The diversity of lineage group I sequences (Table 1) indicates an ancient origin of the region V deleted clade. Furthermore, the highest diversity of group I subgroups and also lineage groups is found in Island Southeast Asia, not Taiwan. Although Taiwan has a high frequency of region V deleted sequences (36%, Table 2), these lineages are restricted to subgroups I.2, I.3, and I.5 like those found in China and Vietnam (Tables 1 and 2).

Generally, the mtDNA data is consistent with the "Eden in the East" scenario. As predicted, Island Southeast Asia contains the highest mtDNA diversity, both of lineage groups and group I subgroups. The high mtDNA diversity within Island Southeast Asia is more consistent with Sundaland as the homeland of an ancient Austronesian language family than as a region of recent, intrusive linguistic expansion. Thus, Island Southeast Asia is the best candidate for a proximate homeland of Micronesians and Polynesians. In particular, Island Southeast Asia shares group I with Micronesia and Polynesia, lineage group I.4 with Yap and Palau, and lineage group III with Palau and Polynesia. Thus three genetic markers not found in Taiwan or mainland Asia are found in Island Southeast Asia and either Micronesia or Polynesia (Table 2). As stated, Oppenheimer (1998) draws from a wide range of disciplines to reconstruct prehistory, including some mtDNA studies (Hertzberg et al. 1989; Lum et al 1994; Redd et al. 1995; Sykes et al. 1995). Therefore, although the present mtDNA analysis is consistent with the "Eden in the East" scenario, the two are not completely independent.

Autosomal and Y-Chromosome Polymorphisms

Since other chapters of this volume will cover these topics in detail, only the contrasts between mtDNA and autosomal and Y chromosome results will be touched upon here. In general, studies of autosomal loci have revealed a greater similarity between Remote Oceanic and New Guinean gene pools than mtDNA studies (Serjeantson 1985; O'Shaughnessy et al. 1990). Similarly, genetic distances estimated from autosomal short tandem repeat (STR) variation indicates that Remote Oceanic populations are intermediate between Asian and highland New Guinea populations, which was attributed to male biased gene flow (Lum et al. 1998).

A recent analysis of Y chromosome haplotypes has provided direct insight on male gene flow (Su et al. 2000). The main objective of this study was to evaluate the merits of the "Express Train" hypothesis by comparing Y chromosome haplotypes from Taiwanese, Southeast Asian, and Pacific Island populations. The study revealed that the Y chromosome haplotypes of both Taiwan and Polynesia/Micronesia were both low diversity subsets of Southeast Asia, but largely non-overlapping. This result supports the view of Taiwan as a Southeast Asian isolate rather than origin as suggested by Oppenheimer (1998). Furthermore, one haplotype detected (H17) was restricted to New Guinea and island Melanesia. Interestingly, this distribution is coincident with that of the typical Melanesian morphology. Su et al. (2000) interpreted the absence of H17 from Micronesia and Polynesia as evidence of an Asian origin of their Y-chromosomes. This conclusion is inconsistent with the male biased gene flow inferred from previous studies of mtDNA and austosomal loci (Lum et al. 1998). Although the Y chromosome haplotypes found in Polynesia and Micronesia are a subset of those found in Southeast Asia, it is noteworthy that the majority (~60-70%) are H1 and H5 which are found in similar frequencies in New Guinea and Vanuatu. Table 3 shows genetic distances (Nei's D) estimated among regions calculated from the haplotype frequencies of Su et al. (2000) using Gendist (Felsenstein 1993). The largest distance is between Taiwan and Melanesia, while the smallest distances are among Micronesia, Polynesia, and Melanesia. Although Polynesians and Micronesians lack H17, their haplotype profiles are more similar to Melanesians than Island Southeast Asians.

Conclusions

Comparisons of mtDNA lineages, autosomal loci, and Y-chromosome haplotypes highlight the difference between maternal and paternal genetic affinities and suggest that substantial male gene flow among Remote Oceanic and New Guinean populations has occurred at some time in the past. When and under what circumstances this male biased gene flow occurred, as well as numerous other questions concerning interaction among archipelagos remains to be answered.

Table 3. Y-chromosome genetic distances (Nei's D) between regions

	Taiwan	Micronesia	Polynesia
Taiwan	-	2.09	2.17
Southeast Asia	0.76	0.83	0.85
Island Southeast Asia	0.48	0.54	0.74
Micronesia	2.09	-	0.33
Polynesia	2.17	0.33	-
Melanesia	4.62	0.29	0.64

References

Alkire, W.H. 1965. *Lamotrek Atoll and Inter-Island Socioeconomic Ties*. Urbana and London: The University of Illinois Press.

Alkire, W.H. 1972. *Population Dynamics of Woleai and Lamotrek Atolls*. Unpublished, available at Hamilton Library, Pacific Collection. Manao: University of Hawaii.

Alkire, W.H. 1978. Coral Islanders. IN: Goldschmidt, W. (ed.), *Worlds of Man: Studies in Cultural Ecology*. Arlington Heights: AHM Publishing Corp.

Anderson, S., Bankier, A.T., Barrell, B.G., de Bruijin, MH.L., Coulson, A.R., Drouin, J., Eperon, I.C., et al. 1981. Sequence and organization of the human mitochondrial genome. *Nature* 290: 457-465.

Ballinger, S.W., Schurr, T.G., Torroni, A.T., Gan, Y.Y., Hodge, J.A., Hassan, K., Chen, K-H., and Wallace, D.C. 1992. Southeast Asian mitochondrial DNA analysis reveals genetic continuity of ancient Mongoloid migrations. *Genetics* 130:139-152.

Barratt, G. 1988. Carolinean contacts with the islands of the Marianas: The European record. Micronesian Archaeological Survey Report No. 25. Saipan: Division of Historic Preservation, Department of Community and Cultural Affairs.

Bellwood, P. 1985. *Prehistory of the Indo-Malaysian Archipelago*. Honolulu: University of Hawaii Press.

Bender, B.W. 1971. Micronesian languages. IN: Sebeok, T.A. (ed.), *Current Trends in Linguistics*. Mouton: The Hague, pp. 426-465.

Blust, R. 1981. Variation in retention rate among Austronesian languages. Paper presented at the Third International Conference on Austronesian Linguistics, Bali, 1981.

Blust, R. 1995. The position of Formosan languages: methods and theory in Austronesian comparative linguistics. IN: Li, P. J-K., et al. (eds.), *Austronesian Studies Relating to Taiwan.*. Symposium Series of the Institute of History and Philology, No. 3. Taipei: Academia Sinica, pp. 585-650.

Cann, R.L. and Wilson, A.C. 1983. Length mutations in human mitochondrial DNA. *Genetics* 104:699-711.

Darwin, C. 1859. [1967] *On the Origin of Species by Means of Natural Selection*. New York: Atheneum.

Diamond, J.M. 1988. Express train to Polynesia. *Nature* 336: 307-308.

Diamond, J.M. 2000. Taiwan's gift to the world. *Nature* 403: 709-710.

Felsenstein, J. 1993. PHYLIP (Phylogeny Inference Package) version 3.5c. Distributed by the author. Seattle: Department of Genetics, University of Washington.

Finney, B. 1994. *Voyage of Rediscovery: A Cultural Odyssey Through Polynesia*. Berkeley: University of California Press.

Golson, J. 1977. No room at the top: Agricultural intensification in the New Guinea Highlands. IN: Allen, J., Golson, J., and Jones, R. (eds.), *Sunda and Sahul:*

Finney, B. 1994. *Voyage of Rediscovery: A Cultural Odyssey Through Polynesia*. Berkeley: University of California Press.

Golson, J. 1977. No room at the top: Agricultural intensification in the New Guinea Highlands. IN: Allen, J., Golson, J., and Jones, R. (eds.), *Sunda and Sahul: Prehistoric Studies in Southeast Asia, Melanesia and Australia*. London: Academic Press, pp. 601-638.

Green, R.C. 1991. Near and Remote Oceania: Disestablishing "Melanesia" in cultural history. IN: Pawley, A. (ed.), *Man and a Half: Essays in Pacific Anthropology and Ethnobotony in Honour of Ralph Bulmer*. Auckland: The Polynesian Society, Inc., pp. 491-502.

Green, R.C. 1999. Integrating historical linguistics with archaeology: insights from research in Remote Oceania. *Indo-Pacific Prehistory Association Bull.* 18:3-16.

Groube, L., Chappell, J., Muke, J., and Price, D. 1986. A 40,000 year-old human occupation site at Huon Peninsula, Papua New Guinea. *Nature* 324: 453-455.

Handoko, H.Y., Lum, J.K., Gustiani, Rismalia, Kartapradja, H., Sofro, A.S.M., and Marzuki, S. 2001. Length variation in the COII-tRNA^{Lys} intergenic region of mitochondrial DNA in Indonesian populations. *Hum. Biol.* (in press).

Hertzberg, M., Mickleson, K.N.P., Serjeantson, S.W., Prior, J.F., and Trent, R.J. 1989. An Asian-specific 9-bp deletion of mitochondrial DNA is frequently found in Polynesians. *Am. J. Hum. Genet.* 44: 504-510.

Horai, S., Kondo, R., Nakagawa-Hattori, Y., Hayashi, S., Sonoda, S., and Tajima, K. 1993. Peopling of the Americas, founded by four major lineages of mitochondrial DNA. *Mol. Biol. Evol.* 10: 23-47.

Hunter-Anderson, R., Thompson, G.B., and Moore, D. 1995. Rice as a prehistoric valuable in the Mariana Islands, Micronesia. *Asian Persp.* 34: 69-89.

Intoh, M. 1997. Human dispersal into Micronesia. *Anthropol. Sci.* 105: 15-28.

Irwin, G. 1992. *The Prehistoric Exploration and Colonization of the Pacific*. Cambridge: Cambridge University Press.

Kirch, P.V. 1997. *The Lapita Peoples*. Oxford: Blackwell.

Kirch, P.V. and Hunt, T.L. 1988. Radiocarbon dates from the Mussau Islands and the Lapita colonization of the Southwest Pacific. *R.A.C.A.A.* 30: 161-169.

Lewis, D. 1972. *We, the navigators*. Honolulu: University of Hawaii Press.

Lingenfelter, S.G. 1975. *Yap: Political Leadership and Culture Change in an Island Society*. Honolulu: University of Hawaii Press.

Lorenz, J. and Smith, D.G. 1997. Distribution of sequence variation in the mtDNA control region of Native North Americans. *Hum. Biol.* 69: 749-776.

Lum, J.K. 1998. Central and Eastern Micronesia: Genetics, the overnight voyage, and linguistic divergence. *Man and Culture in Oceania* 14: 69-80.

Lum, J.K. and Cann, R.L. 1998. MtDNA and language support a common origin of Micronesians and Polynesians in island Southeast Asia. *Am. J. Phys. Anthropol.* 105: 109-119.

Lum, J.K. and Cann, R.L. 2000. mtDNA lineage analyses: origins and migrations of Micronesians and Polynesians. *Am. J. Phys. Anthropol.* 113: 151-168.

Lum, J.K., Rickards, O., Ching, C., and Cann, R.L. 1994. Polynesian mitochondrial DNAs reveal three deep maternal lineage clusters. *Hum. Biol.* 66: 567-590.

Lum, J.K., Cann, R.L., Martinson, J.J., and Jorde, L.B. 1998. Mitochondrial and nuclear genetic relationships among Pacific island and Asian populations. *Am. J. Hum. Genet.* 63: 613-624.

Marck, J.C. 1986. Micronesian dialects and the overnight voyage. *J. Polynesian Soc.* 95: 253-258.

Moore, C.C. and Romney, A.K. 1994. Material culture, geographic propinquity, and linguistic affiliation on the north coast of New Guinea: A reanalysis of Welsch, Terrell, and Nadolski (1992). *Am. Anthropol.* 96: 370-396.

O'Shaughnessy, D.F., Hill, A.V.S., Bowden, D.K., Weatherall, D.J., and Clegg, J.B. 1990. Globin genes in Micronesia: origins and affinities of Pacific island peoples. *Am. J. Hum. Genet.* 46: 144-155.

Oppenheimer, S. 1998. *Eden in the East:The Drowned Continent of Southeast Asia.* London: Phoenix.

Pawley, A. and Green, R.C. 1973. Dating the dispersal of the Oceanic languages. *Ocean Ling.* 12: 1-67.

Pietrusewsky, M. 1990a. Craniofacial variation in Micronesia and the Pacific: A multivariate study. *Micronesica Suppl.* 2:373-402.

Pietrusewsky, M. 1990b. Craniofacial variation in Australasian and Pacific populations. *Am. J. Phys. Anthropol.* 82: 319-340.

Redd, A.J., Takezaki, N., Sherry, S.T., McGarvey, S.T., Sofro, A.S.M., and Stoneking, M. 1995. Evolutionary history of the COII/tRNA^{Lys} intergenic 9 base pair deletion in human mitochondrial DNAs from the Pacific. *Mol. Biol. Evol.*. 12:604-615.

Richards, M., Oppenheimer, S., and Sykes, B. 1998. mtDNA suggests Polynesian origins in eastern Indonesia. *Am. J. Hum. Genet.* 63: 1234-1236.

Rickards, O., Martinex-Labarga, C., Lum, J.K., De Stefano, G.F., and Cann, R.L. 1999. Mitochondrial DNA history of the Cayapa Amerinds of Ecuador: Detection of additional founding lineages for the native American populations. *Am. J. Hum. Genet.*. 65: 519-530.

Ross, M. 1996. Is Yapese Oceanic? IN: Nothofer, B. (ed.), *Reconstruction, Classification, Description. Festschrift in Honor of Isidore Dyen.* Hamburg: Verlag Meyer and Co., pp. 121-166.

Serjeantson, S.W. 1985. Migration and admixture in the Pacific: insights provided by Human Leucocyte Antigens. IN: Kirk, R. and Szathmary, E. (eds.), *Out of Asia: Peopling the Americas and the Pacific.* Canberra: The Journal of Pacific History Inc., pp. 133-145.

Soodyall, H., Vigilant, L., Hill, A.V., Stoneking, M., and Jenkins, T. 1996. mtDNA control-region sequence variation suggests multiple independent origins of an "Asian-specific" 9-bp deletion in sub-Saharan Africans. *Am. J. Hum. Genet.* 58: 595-608.

Spriggs, M. 1989. The dating of the Island Southeast Asia Neolithic: an attempt at chronometric hygiene and linguistic correlation. *Antiq.* 63: 587-613.

Su, B., Jin, L., Underhill, P., Martinson, J., Saha, N., McGarvey, S.T., Shriver, M.D., Chu. J., Oefner, P., Chakraborty, R., and Deka, R. 2000. Polynesian origins: Insights from the Y chromosome. *Proc. Natl. Acad. Sci.* USA 97: 8225-8228.

Sykes, B., Leiboff, A., Low-Beer, J., Tetzner, S., and Richards, M. 1995. The origins of the Polynesians: and interpretation from mitochondrial lineage analysis. *Am. J. Hum. Genet.* 57: 1463-1475.

Terrell, J.E. 1988. History as a family tree, history as an entangled bank: constructing images and interpretations of prehistory in the South Pacific. *Antiq.* 62: 642-657.

Terrell, J.E., Hunt, T.L., and Gosden, C. 1997. The dimensions of social life in the Pacific. *Cur. Anthropol.* 38: 155-195.

Terrell, J.E. and Welsch, R.L. 1997. Lapita and the temporal geography of prehistory. *Antiq.* 71: 548-572.

Terrell, J.E., Kelly, K.M. and Rainbird, P. 2001. Forgone conclusions? An analysis of the concepts of "Austronesians" and "Papuans". *Cur. Anthropol.* (in press).

Torroni, A., Petrozzi, M., Santolamazza, P., Sellitto, D., Cruciani, F., and Scozzari, R. 1995. About the "Asian"-specific 9-bp deletion of mtDNA. *Am. J. Hum. Genet.* 57: 507-508.

Torroni, A., Schurr, T.G., Cabell, M.F., Brown, M.D., Neel, J.V., Larsen, M., Smith, D.G., et al. 1993. Asian affinities and continental radiation of the four founding Native American mtDNAs. *Am. J. Hum. Genet.* 53: 563-90.

Vigilant, L., Pennington, R., Harpending, H., Kocher, T.D., and Wilson, A.C. 1989. Mitochondrial DNA sequences in single hairs from a southern African population. *Proc. Natl. Acad. Sci. USA* 86: 9350-9354.

Vigilant, L., Stoneking, M., Harpending, H., Hawkes, K., and Wilson, A.C. 1991. African populations and the evolution of human mitochondrial DNA. *Science* 253: 1503-1507.

Watkins, W.S., Bamshad, M., Dixon, M.E., Bhaskara Rao, B., Naidu ,J.M., Reddy, P.G., Prasad B.V.R. et al. 1999. Multiple origins of the mtDNA 9-bp deletion in populations of South India. *Am. J. Phys. Anthropol.* 109: 147-158.

Welsch, R.L., Terrell, J., and Nadolski, J.A. 1992. Language and culture on the north coast of New Guinea. *Am. Anthropol.* 94: 568-600.

White, J.P. and O'Connell, J.F. 1982. *A Prehistory of Australia, New Guinea, and Sahul.* Sydney: Academic Press.

Wickler, S. and Spriggs, M. 1988. Pleistocene human occupation of the Solomon Islands, Melanesia. *Antiq.* 62: 703-706.

Wrischnik, L.A., Higuchi, R.H., Stoneking, M., Erlich, H.A., Arnheim, N., and Wilson, AC. 1987. Length mutations in human mitochondrial DNA: direct sequencing of enzymatically amplified DNA. *Nucl. Acid. Res.* 15: 529-542.

Wurm, S.A. and Hattori, S. (eds.) 1981. *Language Atlas of the Pacific Area.* Canberra: The Australian Academy of the Humanities in collaboration with the Japan Academy.

INDEX

www.ingramcontent.com/pod-product-compliance
Ingram Content Group UK Ltd.
Pitfield, Milton Keynes, MK11 3LW, UK
UKHW022103170926
8624IPUK00024B/33

9 789810 247843